Historical Geography and Geosciences

This book series serves as a broad platform for contributions in the field of Historical Geography and related Geoscience areas. The series welcomes proposals on the history and dynamics of place and space and their influence on past, present and future geographies including historical GIS, cartography and mapping, climatology, climate history, meteorology and atmospheric sciences, environmental geography, hydrology, geology, oceanography, water management, instrumentation, geographical traditions, historical geography of urban areas, settlements and landscapes, historical regional studies, history of geography and historic geographers and geoscientists among other topically related areas and other interdisciplinary approaches. Contributions on past (extreme) weather events or natural disasters including regional and global reanalysis studies also fit into the series.

Publishing a broad portfolio of peer-reviewed scientific books Historical Geography and Geosciences contains research monographs, edited volumes, advanced and undergraduate level textbooks, as well as conference proceedings. This series appeals to scientists, practitioners and students in the fields of geography and history as well as related disciplines, with exceptional titles that are attractive to a popular science audience.

If you are interested in contributing to this book series, please contact the Publisher.

More information about this series at https://link.springer.com/bookseries/15611

Jacobo García-Álvarez •
Paloma Puente-Lozano
Editors

Beneath the Lines

Borders and Boundary-Making from the 18th to the 20th Century

Springer

Editors
Jacobo García-Álvarez
Department of Humanities: History, Geography and Art
Carlos III University of Madrid
Getafe, Madrid, Spain

Paloma Puente-Lozano
Department of Humanities: History, Geography and Art
Carlos III University of Madrid
Getafe, Madrid, Spain

ISSN 2520-1379 ISSN 2520-1387 (electronic)
Historical Geography and Geosciences
ISBN 978-3-030-96903-5 ISBN 978-3-030-96904-2 (eBook)
https://doi.org/10.1007/978-3-030-96904-2

This Springer imprint is published by the registered company Springer Nature Switzerland AG
The registered company address is: Gewerbestrasse 11, 6330 Cham, Switzerland

Preface

Beneath the Lines brings together a number of empirically rich and theoretically informed contributions that aim to clarify both geo-historical specificities and common transnational and global features of the modern transformation of political territoriality. The book focuses on the cultures and practices of boundary making that, from the eighteenth century onwards, differently shaped the understanding of states, local populations, and the patterns of borderlands, hence playing a historically prominent role in the territorial production of borders. Why, where and how the complex bundle of doctrines, practices and technologies of boundary making became integral to the eventual accomplishment of modern territorialization of sovereignty are some of the questions that lie at the heart of this book. By making empirical contributions speak to wider ongoing discussions on territory, sovereignty and borders within the framework of recent spatial and global turns, this work adopts an integrative agenda that seeks to strengthen the links between renewed geo-historical studies and more contemporary-oriented border studies, as these fields have tended to eschew each other in recent decades.

The book includes ten contributions by scholars from Spain, France, Italy, Argentina, Brazil and Mexico, and has two main parts. Part I (Chaps. 2–5) provides in-depth analyses of different solutions offered by boundary treaties and the particular workings of boundary commissions when coping with complex realities on the ground so that states could achieve workable and lasting territorial arrangements within borderlands. Part II (Chaps. 6–9) expands previous analyses of borders by connecting them to larger, transversal questions, such as the rather ambiguous and controversial role played by natural borders in boundary making, or the centrality of cartographic practices and other forms of geographical and scientific knowledge to this process. Overall, the first part of this volume highlights the complex interplay of the theories and practices of boundary delimitation, whereas the second sheds light on the intertwined set of cartographic, geographic and political practices that shaped borders and played an essential role in settling territorial conflicts, social unrest and divergent claims over borderlands. In addition to these parts, the introductory chapter provides an extensive overview of recent scholarship about boundary making and modern regimes of territorialization. There is also a concluding essay (Chap. 10), which analyses the contribution of global history to the study of borders and border-making processes.

Throughout the book, each case study pays close attention to primary sources, fleshing out many of the conceptual and methodological claims raised in recent historiographical discussions about global, transnational and bordered spaces, and fully displaying their heuristic potential. Lastly, this book offers the outcomes and ongoing efforts of an extensive body of literature originally written in different languages (French, Spanish, Portuguese, German and Italian) to an English-speaking audience. To date, this body of literature has not only been unavailable to many border scholars, but also on occasions mischaracterized in mainstream border studies. As a whole, the book seeks to convey something of the complexity and richness of different borderland cultures as represented in their shifting territorial patterns, their fluid transnational character, and their highly contested, controversial nature.

The edition of this book and the elaboration of several of the contributions to it (namely Chaps. 1–3, 6 and 7) have been supported by the Spanish Ministry of Economy, Industry and Competitiveness and the European Regional Development Fund (Research Project CSO2015-65301-P, MINECO/FEDER). Chapters 1, 2 and 7 have also been supported by the Spanish State Research Agency (Research Project PID2020-114088GB-I00/AEI/10.13039/501100011033).

Getafe, Spain

Jacobo García-Álvarez
Paloma Puente-Lozano

Contents

Editors and Contributors

About the Editors

Jacobo García-Álvarez is Professor of Human Geography at the Carlos III University of Madrid, which he joined in 2003. His research work has been focused on the fields of history of geography, regional geography, historical geography and political geography, with special attention to politico-administrative divisions, boundary-making processes and territorial identities and ideologies in Spain. He chaired the Commission on the History of Geography within the International Geographical Union and the International Union of History of Science and Technology (2008–2016) and has coordinated the History of Geography, Historical Geography and Political Geography Research Group at the Carlos III University Institute of Historiography since its creation in 2012.

Paloma Puente-Lozano is Lecturer in Human Geography at the Carlos III University of Madrid. She holds a PhD in Humanities (2011) from the same university and has held postdoctoral research and teaching positions at Autonomous University of Barcelona (2011–2014), University Antonio de Nebrija (2015–2017) and Comillas Pontifical University (2016–2017). Her research interests lie at the intersection of historical geography, political philosophy and geopolitics, with special focus on the historical transformations of geographical and political concepts (such as territory, border, sovereignty and place/space). She has extensive international experience as researcher and has published in renowned journals such as *Geoforum*, *Journal of Historical Geography* and *Scripta Nova*.

Contributors

Laura Di Fiore University of Naples Federico II, Naples, Italy

Benjamin Duinat University of Pau and Pays de L'Adour, Pau, France

Jacobo García-Álvarez Department of Humanities: History, Geography and Art, Carlos III University of Madrid, Getafe, Madrid, Spain

Sergio Nunes-Pereira Fluminense Federal University, Niterói, Brazil

Paloma Puente-Lozano Department of Humanities: History, Geography and Art, Carlos III University of Madrid, Getafe, Madrid, Spain

Jean-Yves Puyo Laboratory TREE, UMR 6031, University of Pau and Pays de l'Adour, Pau, France

Cristina Santos-Sánchez Carlos III University of Madrid, Getafe, Spain

Luz María Oralia Tamayo National Autonomous University of Mexico, Mexico City, Mexico

Luis Urteaga University of Barcelona, Barcelona, Spain

Benoit Vaillot University of Toulouse – Jean Jaurès, Toulouse, France; University of Strasbourg, Strasbourg, France

Perla Zusman CONICET and University of Buenos Aires, Buenos Aires, Argentina

Drawing the Lines, Knowing the Borders: An Overview of Recent Scholarship on Boundary-Making and Modern Regimes of Territorialization

1

Paloma Puente-Lozano and Jacobo García-Álvarez

Abstract

This chapter offers a wide overview of the recent scholarship on modern boundary-making in different languages and cultural contexts. It aims to bring together the main outcomes of empirical research and the conceptual issues and theoretical debates having arisen both in the context of the so-called spatial and global turns (e.g. new spatially oriented forms of research in global or transnational history) and in border studies over the last decades. The chapter first elaborates on this wider intellectual background; then highlights the specific contribution made in the book (thematic, methodological and historiographical); and, lastly, introduces the chief insights of each chapter of the book.

This work has been supported by the Spanish Ministry of Economy, Industry and Competitiveness and the European Regional Development Fund (Research Project CSO2015-65301-P, MINECO/FEDER), as well as by the Spanish State Research Agency (Research Project PID2020-114088GB-I00/AEI/10.13039/501100011033).

P. Puente-Lozano (✉) · J. García-Álvarez
Department of Humanities: History, Geography and Art, Carlos III University of Madrid, Getafe, Madrid, Spain
e-mail: ppuente@hum.uc3m.es

J. García-Álvarez
e-mail: jacobo.garcia@uc3m.es

1.1 Geohistoricizing Modern Cultures and Practices of Boundary-Making

Over the last three decades, there has been a noteworthy improvement in knowledge about boundary-making processes and the global territorial shaping of modern states between the eighteenth and the twentieth centuries (Maier 2016; Readman et al. 2014). This has entailed remarkable methodological adjustments, a noticeable widening of primary sources and research materials, and fresh insights skilfully nuancing conventional narratives on borders and state-building, in particular the traditional understanding of boundary lines as the result of mere institutional imposition by central state agencies. Historians, geographers, anthropologists, political scientists, and international relations scholars have carried out in-depth and lengthy analyses on the historical and territorial processes that forged current boundaries in different geographical areas[1].

[1]Just as an example, the following works in different languages might be referred to regarding specific areas. For different European cases, see: Stoklosa and Besier (2014); for Italy: Salvaticci (2005), Meriggi (2016), Di Fiore and Mariggi (2013); for Switzerland: Schroeter (2007); for France: Sahlins (1989, 1998), Sermet (1983), Arvizu (1992, 1997, 2001), Desplat (2002), Capvevila (2009), Catalá et al. (2011), Puyo and García-Álvarez (2016), García-Álvarez and Puyo (2019); for the Portuguese–Spanish border: Garcia (1996), Dias (2009), Godinho (2011), Herzog (2015), García-Álvarez and Puente-Lozano

J. García-Álvarez and P. Puente-Lozano (eds.), *Beneath the Lines*, Historical Geography and Geosciences, https://doi.org/10.1007/978-3-030-96904-2_1

This convergence of different fields and the consequent opening up of compelling new avenues for research epitomize a reinvigoration of studies on boundaries, borders, and borderlands. Significantly, bordering processes have become a thriving object of inquiry no longer confined within the circles of various regional and historical specializations and their rather empirical approaches, nor exclusively tied to contemporary epistemological concerns and political anxieties around globalization and emerging new structures of spatiality (Diener and Hagen 2009).

Even though this is a quite scattered and diverse scholarship, and although inquiry on boundary-making in different regions has hinged around specific research questions and agendas, the literature focused on European borders (and thus also including those of colonial territories) has been described as 'distinctly different' (Readman et al. 2014, p. 4). This recent scholarship has been characterized by a number of common features, such as close attention to a wide variety of actors, interests, and sites, whose complex interactions on the ground were key to bringing about new territorial configurations within borderlands (see di Fiore 2017 for a European overview). Special attention has been paid therein to boundary treaties, widely interpreted as complex and pragmatic tools of territorial management of border areas (García-Álvarez and Puente-Lozano 2017a) and to the multi-faceted role played by international joint boundary commissions within overall boundary-making processes (Donaldson 2008). Both elements (boundary treaties and joint boundary commissions) have been integral to the development and accomplishment of a modern conception of borders and sovereignty, whose territorial understanding in turn lies at the heart of the modern geopolitical imagination (Agnew 2003, p. 2; Galmen 2008). Accordingly, the close link between boundary-making and the political configuration of the modern inter-state system has been in the spotlight in many of these works. As contemporary concerns about shifting patterns of territorialization and bordering have gathered momentum, historical inquiry on the key components and processes that brought about the so-called modern territorial state, which continues to be a prevalent political unit organizing the international system, has likewise gained traction (Agnew 2009; Axtmann 2004; Elden, 2010, 2013b; Sassen 2013).

As a matter of fact, the rapidly shifting geopolitical scenario has raised concerns among scholars about the entangled relations between sovereignty, territory, and borders. This has led them to highlight the negative consequences of a *presentist* understanding that conflates these three elements, since a contemporary overlapping of the said items has entailed misconceptions about their actual historical deployment. Therefore, many scholars have pointed to the need to accurately trace their specific genealogies, as they did not evolve simultaneously and different sets of actors were involved in these processes (Elden 2005, 2013a; Agnew 2005; Shah 2011). It is against this backdrop that further claims have been made that in order to fully explain the origins and transformations of territorial sovereignty, specific attention should be paid to the particular history of the global linearization of borders—i.e. the process by which borders were turned into accurately traced lines (whether agreed in international treaties, drawn on maps, or marked on the ground on the basis of geodetic coordinates)—which came to be assumed globally as the ideal

(2015a, 2015b, 2017a), García-Álvarez (2019); for Germany: Demandt (1990), Dunlop (2015), Polak-Springer (2015), Kulczycki (2016); for Central and Eastern Europe: Bjork (2008), Evans (2006), Bartov and Weitz (2013), Murdock (2011); for Northern maritime borders: Morieux (2016); for the French Flanders: Baycroft (2004); for Eurasian imperial borderlands: Rieber (2014), Barret (1999).For colonial cases, see: for Africa, Blais (2011, 2014), African Union Border Programme (2014), Lefebvre (2011), Sandouno (2014); for the Middle East: Ates (2013); for New Zealand: Byrnes (2001).For Latin America, see: for Mexico, Rebert (2001) and Weber (1992); for Brazil, Langfur (2006); for the southwestern border of the United States, Mandrini (2006), Mandrini and Paz (2003). For North America, see: Johnson and Graybill (2010), Hämäläinen and Johnson (2012); for the southern border with the Spanish Empire in particular: Adelman and Aron (1999), Aron (2006), Blackhawk (2006), Guy and Sheridan (1998), de la Teja and Frank (2005) and Brooks (2002). For Asia, see: Schendel (2005), Misra (2011), Breyfogle (2005).

form of territorial demarcation (Goettlich 2019a, 2019b).

While fixed exclusive territoriality has played a central role in accounts of modern sovereignty, both in historical materialist explanations (Krassner 1999; Teschke 2003; Tilly 1992) and in rather ideational approaches (Ruggie 1993; Bartleson 1995; Philipott, 2001; Strandsbjerg 2010), boundary-making has received much less attention within this scholarship on modern statehood. It is worth noting that the historical and geographical intricacies and variations of the blurred transition from a feudal patrimonial form of political authority to a modern one—that is, a move away from medieval heteronomy and its complicated pattern of interrelated jurisdiction towards modern clear-cut delimited sovereignty—have been extremely helpful for shedding further light on the well-worn 'Westphalian myth' (Osiander 2001; Hobson 2009; Teschke 2003), insofar as both forms of rule overlapped and were often practised simultaneously. Accordingly, critical revisions of imperial and global territorial formations have likewise put the emphasis on the different local instantiations of sovereignty-related processes by revealing their own internal heterogeneity and the uneven exercise of political authority, not least regarding its legal dimension, as stressed by Benton (2010) and many others (Burbank and Cooper 2011; Antunes and Polónia 2016; Benton and Ross 2013). As part of this challenge to conventional narratives about modern states, it has been argued that colonial spaces were central to the genesis and complex evolution of sovereign and territorial statehood, rather than mere passive laboratories where European ideas and models were tested, applied, or imposed. For, as Jordan Branch has put it, 'the character of the modern system of territorially sovereign states resulted from a complex interaction between European colonizing polities and events, actors, and spaces in other parts of the globe' (Branch 2011, p. 277).

Colonial spaces thus became active laboratories where certain political, legal, and spatial practices and ideas first arose in early modernity through the European actors operating there, only later being implemented in metropolitan spaces. These practices and ideas then gathered force through postcolonial independence, with sovereign territorial rule buttressing the 'legitimacy of the new states' (Schulz 2019, p. 879; Lois 2019).

Nevertheless, the important role that boundary-making played therein still remains only partially explored, despite the important fact that scientifically and clear-cut demarcated boundaries actually led to the eventual consolidation of the centuries-long shaping of modern statehood. In this sense, and drawing on previous efforts made along similar lines (Donaldson 2008; Readman et al. 2014, Stokosa and Besier 2014), the empirical contributions in this book seek to shed light on these often overlooked connections between the linearization of borders and the modern production of territory, and so to further contribute to ongoing debates on sovereignty, territoriality, and state-making. By describing modern boundary-making as a process of 'geometric purification' of state territory that was key to the enforcement of a fully fledged spatialized form of sovereignty, this book seeks to give a wider context to this process through a thorough examination of the complex interplay of actors, institutions, forms of knowledge, acts of resistance, and interests that converged in boundary-making practices. Accordingly, the book attempts to highlight the extent to which the worldwide process of linearization of borders actually relied on the steady formation of cultures of territorial demarcation in Europe and in colonial spaces that, despite presenting their own particularities and taking shape at a different pace in different countries, shared remarkable common features and were inextricably connected to state-making practices. It is our contention that carefully analysing delimitation and demarcation practices, i.e. looking 'beneath the lines' as they appear in maps, treaties, and land, is essential if we are to explain why and how linear borders came to be legitimated as a key component of the modern geopolitical imagination and thus of how exclusive territorial rule came to be the norm across the whole international system (Sheehan 2006, p. 6; Ford 2010).

Therefore, the empirically rich and theoretically informed contributions gathered together in this book aim to help clarify the geohistorical specificities and common transnational and global features of the modern transformation of political territoriality. We consider how all this came about through different *cultures and practices of boundary-making* that shaped both states' and local populations' understandings of borders. Those cultures and practices include the ways in which boundary commissions were formed and how they acted and related to local communities; how these communities could legally defend or claim possession of disputed lands; the various and usually context-specific criteria used to decide how and where to draw the boundary line; and how to better resolve territorial issues in borderlands in order to put a definitive end to long-standing social unrest in these areas. All in all, the book attempts to advance the understanding of why, where, and how this complex bundle of doctrines, practices, and technologies of boundary-making became integral to the accomplishment of the modern territorialization of sovereignty.

1.2 The Intellectual Contribution of the Book

Along with the general goal of making these empirical contributions speak to wider ongoing discussions on territory, sovereignty, and borders, this book has adopted an integrative approach, seeking to strengthen the links (and to close the existing gaps, where feasible and desirable) between renewed geohistorical studies, such as that of Herzog (2015, pp. 1–15), and more contemporary-oriented border studies, which are deeply anchored in post-structuralist and post-modern approaches and aim to reinvigorate current theoretical discussions (Parker and Vaughan-Williams 2013; Paasi 2011). Accordingly, this book situates itself in the broader intellectual landscape shaped by the contemporary transformations produced across the humanities and the social sciences by the so-called spatial turn (Warf and Arias 2008; Döring and Thielmann 2008) and the heated controversies that followed it (Lossau and Lippuner 2004; Jerram 2013).

Undoubtedly, recent decades have seen a growing awareness of geographical particularity and the spatial variation of different phenomena across contexts, leading to increased criticism and questioning of the primacy of national perspectives and correlated historiographical approaches structured around the nation (Faist 2012; Amelina et al. 2012). As postcolonial, cultural, and global studies have shifted the focus towards comparisons, connections, flows, transfers, circulations, networks, and relations across time and space (Freist 2017; Middell and Naumann 2010), transnational approaches have been put forth in order to come to terms with the complex spatial nature of many modern phenomena and the historically shifting patterns through which they have found expression (Alcalde 2018; Müller and Torp 2009). Whether moving across different scales of analysis in order to explain actors' embeddedness in multiple contexts (Revel 1996) or adopting strong constructivist notions of space that tend to deconstruct the very idea of scale and explain it as a product of historical processes taking place at different levels (De Vito 2019; Campbell 2018), recent scholarship on global and transnational history has likewise made inroads into deeply 'spatialized' approaches (O'Brien 2006; Patel 2015).

Although often overshadowing (if not ignoring) valuable advances made in the 1970s and 1980s along similar lines, this recent scholarship has not only helped scholars avoid the various 'territorial traps' (Agnew 1994) so well known to political geographers—i.e. taking the nation-state as a given unit of historical and political analysis—but it also has made it possible to overcome previous dualistic visions or approaches that were overly dependent on rigid dialectic relations (e.g. core–periphery, top–down, local–global, or local–national) that tended to oversimplify the complex dynamics of nation-state-making. Converging with the claim of geohistorical studies that multi-scalar and actor-centred approaches are needed to account for the complexities of bordering practices and borderlands (Di Fiore 2015), more nuanced and non-essentialized understandings of historical space have emerged in transnational or global historical studies too, ultimately highlighting a relational and networked sense of spatiality (Conrad 2016, p. 75; Sachsenmaier 2012, p. 81). As transnational, global, or planetary spatial categories or metageographies, such as oceans, empires, continents, civilizations, or worldwide trade networks, have figured prominently in recent scholarship, a heightened sense of their intrinsically unstable nature has also emerged, accounting for the historically fluctuating spatial borders and geographical patterns that shaped these historical phenomena (Lewis and Wigen 1997). The need to study *in situ* the very production of these global processes has in turn led to an increasing cross-fertilization of micro-historical analysis and interactionist methodologies (Bernat and Callafat 2018; De Vries 2019), which has foregrounded the involvement of a wide variety of groups, practices, and institutions in the production of global or transnational territorialities.

Close attention to the subtleties of various actors' strategies as deployed on the ground, including in particular how strategic convergences, negotiations, and tactical accommodations of multiple actors' interests occurred (Duinat 2019), has enabled a richer understanding of political and territorial formations. It has been argued that this highly micro and localized understanding of national or global processes has changed 'our way [of reading] the documents' (Turco 2019, p. 54), since this spatialized or topographical approach to historical sources brings their pragmatism back into the spotlight. By placing primary sources in the wider material and ideational context in which they first emerged, actors' own understandings of these territorial processes are uncovered, and actors' differing perceptions thus come to play a key role in explaining how they became enmeshed within modern stake-making and the production of territorial formations.

The task of integrating these different bodies of literature is taken up here in the spirit of developing a more reflexive stance towards the converging paths of contemporary debates around spatial issues (and, in particular, on borders) by foregrounding their relationship to similar endeavours from the past that are often overlooked, for example, the decades-long efforts of geographers to critically approach spatial units as historically, discursively, and materially constructed (García-Álvarez 2006; Paasi 2020), or the attempt of previous empirical/classical geohistorical studies on boundaries to emphasize the historically fluid, shifting, and networked nature of borderlands (Schofield 2015; Zusman and Barros 2000; Nuñez et al. 2017).

Lastly, before delving properly into the content and the specific theoretical framework of the chapters, it is worth making two further remarks on the overall contribution of the book. Besides the *intellectual contribution* of connecting empirical inquiry on boundary-making with contemporary historiographical and methodological issues developed in other fields—as is the case with discussions on global and transnational history—, it has already been pointed out that a *thematic contribution* is attempted here regarding renewed scholarship on borders, territory, and sovereignty and their entangled genealogies. This book seeks to highlight that, even though the global linearization of borders was strongly tied up in imperial formations (Goettlich 2019b), this did not happen without close consideration of and adaptation to the geographical, social, and historical specificities and conditions of borderlands in different areas, as shown in each of the chapters by accurately researched empirical cases. In this sense, this book also makes a *methodological contribution*, as comparatist and transnational approaches crystallize here in a close and transversal reading of primary sources—that is, one that takes into account the available archival documentation produced by all the actors involved in boundary-making processes, whether international, state, or local, and whether external/exogenous or internal/indigenous. This close attention to primary sources makes it possible for many of the conceptual and methodological claims put forth in rather abstract, theory-heavy historiographical discussions about global or transnational approaches (Patel 2015, p. 6) to be fleshed out and fully displayed in their heuristic potential.

Finally, the book offers to a wide English-speaking audience the results and ongoing efforts of a very abundant literature written in different languages (French, Spanish, Portuguese, German, and Italian), which has until now been unavailable to many border scholars. This is of paramount importance, as European boundary-making practices have sometimes been mischaracterized in mainstream Anglophone border studies because of this lack of accessibility to a multilingual and scattered scholarship, which has been shaped by specific academic traditions and theoretical concerns. By organizing the chapters according to a thematic criterion, and by offering in what follows in this introduction a systematic overview of the common conceptual threads and shared ideas that cut across the chapters, this book seeks to convey something of the complexity and richness of different borderland cultures as instantiated in their shifting territorial patterns, their fluid and transnational character, and their highly contested and controversial nature. Readers will be offered a wide array of case studies focusing, in the first part, on the complex interplay of theories and practices of boundary delimitation and demarcation, and, in the second part, on the intertwined cartographic, geographical, legal, diplomatic, and political practices that shaped borders and that played an essential role in managing and settling territorial conflicts, social unrest, and divergent claims over borderlands.

1.3 Historical Geographies of Borders and Boundary-Making (Eighteenth–Twentieth Centuries): Theoretical Threads and Common Insights in the Chapters

As will be shown in detail in this Chapter, the late modern period was marked by the consolidation and global expansion of the territorial state, characterized by the existence of a sovereign power exercised exclusively over a discrete and continuous territory, accurately delimited by linear borders. Although it is commonly claimed that the diplomatic origins of modern statehood go back as far as the Peace of Westphalia (whose practical significance to modern state-making has been increasingly questioned, as previously mentioned), the shift towards a territorial conception of sovereignty and a linear conception of borders was not effectively enforced in Europe until the French Revolution and, in particular, the Congress of Vienna and the Treaties of Paris of 1815 to 1816. Within the wider context of the subsequent expansion of nationalism and imperialism, the provisions in said treaties—just like those contained in the boundary treaties that proliferated thereafter across Europe—entailed the accomplishment of the centuries-long process of smoothing down the state's territory and making it as homogeneous and integrated as possible.

Unquestionably, the formation of the modern state world map, which as a process cannot be separated from the worldwide adoption and implementation of linear boundaries, occurred at an uneven pace both temporally and spatially. Works by Foucher (1991), Prescott and Triggs (2008), and Keating (2018), among others, have carried out insightful analyses in order to systematize the major patterns and trends in modern boundary-making and state creation. To refer to this process, Foucher coined the term '*horogenèse*', a French neologism based on the Greek term '*hóros*', referring to the landmark used in Ancient Greece to delimit land tenure. Within this *horogenèse*, a series of milestones can be distinguished, such as the most important peace conferences and treaties of the nineteenth and twentieth centuries, as well as certain periods of time marked by what Keating (2018, p. 52) has called 'waves of country creation' (e.g. the period from 1810 to 1830, in which many Latin American states became independent; the years 1910–1923, with the balkanization of the Austro-Hungarian and Ottoman empires; the period 1945–1975, when the decolonization of most African and Asian countries took place; or the years following the fall of the Berlin Wall, which saw the collapse of the multinational federations of the USSR and Yugoslavia). Likewise, the making of the political world map was led by a rather select group of '*grand Etats traceurs*' ('great drawer states'), as Foucher (1991) refers to the European imperial and colonial powers responsible for the shaping of most of the present-day borders across Latin America, Africa, and major parts of Asia. As will be explained below, the preparation and effective execution of modern boundary treaties was entrusted to joint boundary commissions, which in many colonial regions (scarcely known from a geographical point of view) were really organized as border expeditions.

The legalization and purification of the linear border undertaken by these treaties, and by the joint boundary commissions that prepared or executed them, should not, in any case, make us forget the role played by other actors and periods that were key to the making of the border, within the context of long-term processes that in some European countries date back to the Middle and Early Modern Ages. For example, as shown in some recent works (Herzog 2015; Capdevila 2009), and as argued in some of the chapters of this book, the Boundary Treaties of Lisbon (1864 and 1926), between Spain and Portugal, and of Bayonne (1856–1866), between Spain and France, gave international legal status to two linear borders defined de facto, for the most part, by the possession practices developed by local border communities over the previous centuries. Thus, to a great extent, by virtue of these treaties the municipal boundaries between states became fully fledged international borders. In countries with a long political history, the replacement of the soft and diffuse borders (frontiers) characteristic of the jurisdictional state by the hard and linear borders (boundaries) of the territorial state, and hence the effective implementation of the Westphalian order, can only be properly understood on a scale spanning several centuries, in which the liberal revolutions are more a culmination than a disruption.

1.3.1 Boundary Treaties and Joint Commissions: Theoretical Foundations and the Praxis on the Ground

Against the historical backdrop outlined above, Chaps. 2–5 in the first part highlight the flexible and context-sensitive character of modern boundary treaties and foreground the particular practices and dynamics that were integral to the functioning of joint boundary commissions across Europe throughout the eighteenth and nineteenth centuries. This is done by offering a detailed analysis of the solutions and criteria adopted by the Spanish–Portuguese Boundary Treaties of 1864 (Chap. 2) and 1893 (Chap. 3), delimiting a large part of the border between these two countries in the Iberian Peninsula; of the work carried out on the ground between 1871 and 1877 by the Franco–German Joint Boundary Commission to effectively demarcate the new boundary line after the annexation of Alsace-Lorraine by Germany (Chap. 4); and, lastly, of the role of Mexican geographer-engineers who took part in the boundary commissions that delimited Mexican borders in the nineteenth century, both in the north (with the United States) and in the south (with Guatemala and Belize) (Chap. 5).

The contributions in this first part build on the existing literature on boundary treaties and joint commissions, which has brought about new interpretations of how delimitation and demarcation were actually performed through a complex set of practices, interests, technologies, and doctrines that were mobilized throughout these processes. It is worth noting that, in spite of the specificities of each case study, modern boundary treaties shared a number of common goals and features that expressed larger political assumptions proper to both enlightened absolutist and emerging liberal governments (Rowe 2014; Ricuperati 2001). These objectives clearly demonstrate the geopolitical and territorial dimension that boundary treaties uphold, which made them far more than just technical or neutral devices, as is highlighted in particular in Chaps. 2, 3 and 4.

To start with, boundary treaties set out to resolve international disputes over contested territories that had provoked war-like situations; such was the case for the American colonies, as negotiations to end the war included prominently discussions about where the border was to fall (Carroll 2001). Likewise, reinforcing the state's effective control over every corner of its territory was another important goal that boundary treaties helped to achieve (Branch 2013, pp. 120–141). For centuries, borderlands had retained significant degrees of autonomy, as their territorial patterns had traditionally been shaped by local customs and actual forms of use and possession of land (Herzog 2015). This often included smuggling and other illegal transborder practices that resulted in heavy economic losses for state coffers and significant flaws in taxation systems. As part of the drive to centralize power that was inherent to modern statehood, border areas were increasingly made amenable to states' territorial practices and patterns. As conflicts in these territorial fringes continued to worsen throughout the nineteenth century, as is made clear in Chap. 2, borderlands became a key object of state-driven territorial management. Accordingly, boundary treaties firmly attempted to put an end to long-standing territorial disputes in borderlands, which in many cases entailed violent conflicts and significant disturbances among border populations over access to agro-silvo-pastoral resources and local lands, most often shared by those groups (Poujade 1998; Dornel 2002; García-Álvarez and Puyo 2015). In fact, the reference to social unrest and territorial issues permeates the rhetoric of boundary treaties, and it is an element that is systematically conveyed therein: it was claimed in boundary treaties that accurately fixing the boundary line and eliminating all remaining uncertainties and vagueness in its layout was integral for achieving the 'pacification' of borderlands. Defining borders 'with the utmost distinction and clarity' (as stated in Article 22 of the 1750 Treaty of Madrid, which aimed to put an end to border conflicts between the Spanish and Portuguese empires in South America) or even 'with full accuracy and precision' (preamble of the 1856 Treaty of Bayonne) was held up as a necessary condition for resolving the recurring border disputes and for ensuring harmony and good neighbourly relations, and thus for bringing about lasting stability.

Yet, bringing full clarity to borderlands frequently entailed the suppression of non-territorial structures of authority (those linked to feudal and jurisdictional rights or privileges) and of enclaves and spaces of unclear or shared sovereignty, which were so common in the Ancien Régime and which were generously interspersed along borders. Within the political culture underlying modern boundary treaties, these jurisdictionally ambiguous spaces had to be eliminated, exchanged, or divided between states so that they could be clearly integrated into one or the other (Bloomley 2003). What these particular practices revealed was that when the assertion of the state's political authority was at stake in certain border areas, the full accomplishment of modern territorial statehood went hand in hand with the linearization of borders, upon which it was dependent. Boundary treaties helped central agencies to ensure the long-term strengthening of state sovereignty.

Certainly, scholarship on boundary-making has emphasized this aspect of boundary commissions as 'crucial catalysts' (Rankin 2008, p. 422) in defining, establishing, and entrenching the state's territorial framework, regarding both colonial policies overseas and domestic ones. Similarly, in order to fully account for their complex and flexible nature, it is important to bear in mind the wide variety of situations and territorial issues that boundary treaties addressed, ranging from the activities allowed in border areas to the regulations necessary to prevent smuggling and other illegal practices. This made boundary treaties highly pragmatic and flexible instruments oriented towards practical goals, attempting to settle disputes over land, the use of resources, and rights at the local level (García-Álvarez and Puente-Lozano 2017b).

Putting to work this on-the-ground approach with regard to boundary-making is important because boundary treaties frequently conveyed rich understandings of borderlands as dynamic and fluid spaces (Zusman 2006)—a fact frequently overlooked in the literature on state-making and modern territoriality. The eclectic, pragmatic, and empirical criteria embraced and the wide variety of solutions adopted in boundary treaties (García-Álvarez and Puyo 2015) have increasingly called into question mainstream interpretations of state-driven territorial demarcation (as featured in Craib 2004; Carter, 2007; Sletto 2001).

Of particular importance here is the fact that the very functioning of the boundary commissions that were established to prepare, negotiate, and effectively execute boundary treaties enabled the emergence of this eclecticism, empiricism, and context-sensitive outlook. This was so because general and abstract principles (whether a tenet of international law or a geographical or physical criterion) tended to be avoided when tackling border issues. For example, drawing systematically on general tenets in order to solve the whole set of different territorial issues, while discussed, was firmly rejected by states. Border disputes were considered on a case-by-case basis, and negotiations and works were carried out in a markedly empiricist and pragmatic way (García-Álvarez and Puente-Lozano 2017b). This is widely shown in the chapters in this part, where a number of delimitation criteria are analysed as discussed and mobilized by different actors and in different stages of the process.

Empiricism and accommodation to local (social, economic, topographical, historical) features were key aspects of boundary treaties because—among other reasons—states were well aware of the fact that for a boundary line to be successful and effective, local populations should be consulted on the process: the ineffectiveness and de facto rejection of legal and diplomatic solutions adopted by central authorities without taking into account local interests had contributed over the centuries to the persistence and entrenchment of territorial issues in borderlands (Arvizu 2001), often leading to fresh disputes and fuelling the conflict, as highlighted in Chaps. 2 and 3.

As recent scholarship has made clear, at least regarding Europe, boundary commissions proved to be key actors throughout boundary-making processes, not least because of the continued interaction with local actors that they made possible when carrying out fieldwork (Rankin 2008; Guy 2008). As far as the cases analysed in this book are concerned, these commissions were composed of diplomats and technical officials, mostly military, who had received sufficient scientific training to perform the geographical and cartographic tasks required for their missions (Capdevila 2009, pp. 57–58; García-Álvarez and Puente-Lozano 2015a). This is so because boundary commissions' work consisted of highly technical tasks and, in each case, the governments involved negotiated and outlined detailed instructions for commissioners to follow.

It is worth highlighting that undertaking these tasks was key for boundary commissions to clarify what *was really at stake* in these territorial issues. In particular, this entailed elucidating the actual needs of the local populations, considered within the broader social and economic context of local practices and customs. Accordingly, joint boundary commissions have been described as *decisive instruments* not only in the territorialization of states' political authority because of how they helped entrench modern practices and ideas of exclusive sovereignty (García-Álvarez 2019; Blais 2011), but also and equally importantly in the creation of an 'institutional space' bringing together the different actors involved in the process, either affected by or calling for a clear and durable boundary line (García-Álvarez and Puente-Lozano 2017a, p. 2).

To the extent that they translated and mediated different actors' multi-scalar concerns, territorial visions, and interests through their important interventions during the whole process of boundary delimitation and demarcation, boundary commissions actually shaped and modified the relations between local populations and governments. They made local claims, customs, and needs comprehensible and acceptable to statesmen and governments, and they contributed to achieving a certain balance between all of the different interests at stake, at least when possible (Donaldson 2008). This can be traced back to a long history of diplomatic negotiations and inter-state will to manage border issues, accounting for centuries-long complex interactions between local populations and central state agencies (Raviola 2010; Mongiano 1991; Chavarría 2005; Fratini 2003).

Within the framework of wider contributions, in Chap. 2 Jacobo García-Álvarez and Paloma Puente-Lozano provide an interpretation of the 1864 Treaty of Lisbon, which delimited the boundary between Spain and Portugal from the mouth of the Minho River to the confluence of the rivers Caia and Guadiana, as a paradigmatic example of how modern ideas of territorial sovereignty were fully implemented through a series of operations and practices that, as a whole, implied a sort of 'geometric and geopolitical purification of the territory', as the authors have put it. The chapter argues that, even though the Treaty of Lisbon did not 'create' the boundary line as such, it crucially gave it its international legal status and, more importantly, helped to achieve the centuries-long transformation of state territory into a continuous, homogeneous, and compact space. This sort of 'geometric purification' entailed the erasure of the spatial impurities or 'irregularities' (such as enclaves, lands of disputed, shared, or dubious ownership, villages divided by the boundary line, and territories of special status that in many respects circumvented the sovereignty of both kingdoms) that were not only the root of many territorial disputes and much social unrest among border communities, but also the material and legal impediment preventing a state sovereignty from being fully effective in borderlands.

As a further exemplification of this drive to 'purify' territory, in Chap. 3 Cristina Santos-Sánchez addresses a specific conflict over shared lands in the so-called La Contienda, an area located on the border between Spain and Portugal that had been in common use over the centuries by the Spanish towns of Aroche and Encinasola and by the Portuguese town of Moura. The common use dated back to 1542, when the first demarcation called *La Concordata* was concluded between these three localities, leaving the most disputed lands undivided. The chapter offers a telling example of the centuries-long dynamics that gave shape to borders, and the extent to which the increasing disputes over lands throughout the eighteenth and nineteenth centuries made the situation untenable, with subsequent attempted solutions falling short. Hence, new criteria and arrangements were needed. To this end, close and detailed knowledge of these lands was likewise required, and boundary commissions thus took full charge. By analysing the origins and the evolution of the conflict and explaining how it eventually came to be resolved through the 1893 *Convenio de Límites entre España y Portugal en la Dehesa de 'La Contienda'*, the chapter makes clear that in order for the territory to be effectively ordered and managed, clear-cut divisions and indisputable property were needed, just as an equitable and workable distribution of the lands among the major actors involved in the conflict was key to ensuring that the boundary line was accepted and enforced.

In Chap. 4, Benoit Vaillot focuses on the issues addressed by the boundary commission that undertook the demarcation on the ground of the Franco–German border after the Franco–Prussian War of 1870. The author posits that the genesis of this border cannot be reduced to the simple drawing of the boundary line on the small-scale map that was attached to the 1871 Treaty of Frankfurt putting an end to the war. Instead, attention should be paid to the fact that it took no less than seven years for the border to be traced in the field, which involved many different actors and some important changes to the original layout of the boundary. After foregrounding the conflicting rationales (ethnolinguistic, strategic, and economic) underlying the delineation of the border in the said Treaty, the author provides a thorough analysis of the activities of the Franco–German Joint Boundary Commission, drawing on rich archival documents and other materials that make it possible to re-create an 'insider perspective'. By examining two of the major territorial issues that the Commission had to tackle (in the so-called Two Raons and Avricourt), the chapter makes important inroads into a 'tangible understanding' of the material production of the border that underlines important examples of the local population's agency.

Finally, in Chap. 5 Luz María Tamayo offers an overview of the delimitation of Mexican borders in the nineteenth century, both in the north (with the United States) and in the south (with Guatemala and Belize), by closely analysing boundary commissions' work and the relationships and inequalities between counterparts.

Significantly, Mexican geographer-engineers undertook this task because they possessed the knowledge of the astronomical and topographical methods necessary to position permanent boundary markers whose coordinates (latitude and longitude) could be identified on a map and therefore be acknowledged as Mexico's international borders. This chapter affirms the 'geographical value' of these often-unacknowledged tasks, and it accordingly attempts to reveal the details and complexities of the commissioners' work, along with the numerous obstacles, shortages, deficiencies, or even diseases or deaths of fellow expedition members that they had to cope with. All in all, the chapter shows that not only did the Mexican commissioners keep up with the US engineers' expertise, technologies, and political support, but, more importantly, that the Exploratory Geographical Commission that was created on the basis of their work was responsible for the best maps of the country available until the mid-twentieth century, giving a lasting and accurate delineation of Mexico's territory.

1.3.2 Contesting the Naturalness of Borders: Cartographic Practices, Geographical Knowledge, and Territorial Issues

Chapters 6, 7, 8 and 9 in the second part of the book expand on previous analyses of boundary treaties and boundary commissions by connecting them to a couple of broader and transversal questions: first, the rather ambiguous and controversial role played by natural borders (such as mountain ranges and rivers) in boundary-making; and, second, the centrality of cartographic practices and other forms of geographical and scientific knowledge to these kinds of processes. Surveying, mapping, and other cartography-related practices have proved to be essential in boundary-making, not only by accurately representing the actual location of the boundary line, but also by making it possible to come to terms with the very nature of the issues at stake in many border conflicts and, consequently, helping the parties to flesh out their arguments and to reach workable agreements.

Throughout this second part, the claim is made that looking very closely at these practices and technologies as displayed and performed on the ground—what, following French scholar Isabelle Laboulais (2008), has been called looking 'behind the scenes' in Chap. 8—is a fundamental methodological advancement that sheds new light on the multifold material assemblage effectively allowing modern borders to come into being. Accordingly, illustrating the 'artificial' character of borders, whether they are based on natural features or not, is a major theme of this part, as it is of the entire book, given the emphasis that has been placed on geohistoricizing boundary-making analysis. In this sense, the empirical contributions in the second part build on the existing literature, which has, on the one hand, called into question the 'enduring geographical myth of natural borders' (Fall 2010), and, on the other hand, argued that territory and state-making are to be approached as a heterogeneous socio-technical assemblage in which cartographic practices loom large (Branch 2013; Painter 2010; Strandsbjerg 2008).

Insofar as the recent literature has argued that border regions, local populations, and nature itself became enmeshed in broader institutional statist cultures of quantification, scientificity, and efficacy, a number of 'calculative techniques'—called so following Foucauldian governmental analysis (Elden 2007a, 2007b)—have likewise gained traction as a key explanatory element of modern state territoriality. Moreover, this territoriality has been highlighted mainly on the basis of this process of turning nature and heterogeneous places into homogeneous rational 'calculable space' (Rose-Redwood 2012) in order for control and security to be effectively enforced by central state agencies, as is also underlined in the concluding Chapter by Laura di Fiore. This literature has significantly contributed to unveiling the false 'naturalness' of borders and state spatialities and to exposing the territorial ideologies underlying the comprehensive legal frameworks enabling state-driven practices (Brighenti 2006; Hannah 2000). Consequently, technology-dependent practices, such as census-taking, statistical and political arithmetic, surveying, and map-making, among others regarded as 'general sciences of order', as Hannah (2009, p. 66) has put it, are claimed to have performed a key role in the modern production of territory (Crampton 2010; Branch 2017). More significantly, it has been shown that these practices were intrinsically entangled within broader *power practices and systems*, insofar as they made both populations and lands amenable to scientific classification, measurement, ordering, and calculation, and hence to governance (Jesoop 2007). It is in this sense that Goettlich (2019a) has talked about 'survey rationality as a mode of territorial governance'.

Against this backdrop, the exercise of sovereignty appears through a Foucauldian lens as a 'right of seizure' (Foucault 1978, p. 136), since counting, owning, distributing, partitioning, mapping, and bordering lands were crucial practices of knowledge and power allowing political authority to be enacted. Among the various scientific practices bringing about this 'epistemological structuring of territory' (Hannah 2009, p. 68), cartography has played a prominent role within recent scholarship.

As a matter of fact, Jordan Branch (2013) has coined the term 'cartographic state' to claim that map-making was key to the emergence of modern territoriality, as the division of space into discretely drawn units was necessary for the shift from a medieval to a modern understanding of political authority. Arguing along the same lines as previous works on this topic (Akerman 1995; Biggs 1999; Black 1997; Steinberg 2005), Branch has posited that boundaries 'appeared first in the representational space of maps and only subsequently in political practices on the ground' (Branch 2013, p. 3). In a similar vein, Jeppe Strandsbjerg (2012, p. 818) has coined the term 'cartopolitics' to describe 'the process where questions pertaining to sovereign control over space are decided through cartography and law'.

Although empirical studies have recently helped to nuance the overstatement of ideational elements and to explain the dynamic back-and-forth interaction between cartography and territory, it is beyond any doubt that maps firmly contributed to the linearization of borders and the purification of state territory, and that this was an important element within the broader drive towards rationalization and calculation that is inherent to modernity (Elden 2006, 2013a). As the chapters in the first part already made clear, scientific cartography and accurate geographical descriptions (drawn up using geometric and topographical parameters) were part of boundary treaties and had a legal nature. Besides the important link between cartography and sovereignty, it is worth stressing the rich and diverse roles that maps played in boundary-making. In this sense, maps and statistics gathered or purposefully developed by joint boundary commissions gave diplomats and statesmen first-hand knowledge of the current state of borderlands—for example, population, surface area, land uses, and the agricultural qualities of the disputed lands. These were essential parameters both for assessing the relevance and strategic nature of the terrains at stake and for understanding the actual needs of the local populations involved in those conflicts, as the chapters in the second part highlight. Dwelling thoroughly on maps, governments were able to quantitatively and qualitatively appraise the territorial exchanges, transfers, or compensations that were offered as a solution to border issues and thus enhanced the centuries-long drive to homogenize state territory.

Cartography as developed during negotiations of boundary treaties and throughout demarcation processes led the boundary to be conceived of as a sharp line separating the sovereign territories of states, as well as firmly integrating borderlands into the wider grid of abstract and homogeneous state spatiality. It is precisely the fact of having states' territories bounded and fixed *in and through* maps which has been claimed to have led to a 'naturalization' of territory, borders, and sovereignty themselves, and to be at the origin of the unquestioned entanglement of these three elements, as mentioned at the beginning of this introduction. Not only have territories frequently been imagined as being naturally separated from one another (as set out in maps as given, discrete spaces), but the fact that natural features have long been used to outline boundaries has also contributed significantly to the reification of borders.

In this sense, it is important to mention that the so-called theory of natural boundaries—that is, the widely held assumption that natural geographical features, such as mountain ranges and large rivers, were suitable elements for the establishment of the borders of political communities—is a rather ancient doctrine and practice across the globe. For example, it was clearly expressed in Roman geography and law (Hekster and Kaizer 2009). As stated in Chap. 7 and already widely addressed and criticized in the literature (Brigham 1919; Brunhes and Vallaux 1921; Zeller 1933; Ancel 1938; Minghi 1963; Sahlins 1990; Fall 2005, Rankin and Schofield 2004; Schaffter et al. 2010), it has traditionally been claimed that natural boundaries offer key strategic properties for state-making, such as easy visualization and control over far-reaching lands, as well as their apparent permanence and durability (Kanska and Manko 2003). As is well known, it was primarily during the seventeenth and eighteenth centuries that the idea of 'natural borders' was given a more solid grounding and was practically applied by modern states. This largely reflected a military, defensive conception of the border and a view of nature as divinely established order (Glacken 1967, pp. 504–654).

Moreover, at that time, military thinking about terrain saw this type of natural boundary (and particularly the lines of mountain ridges and watersheds), as well as the fortifications built in their vicinity, not only as obstacles to conquer but also (or primarily) as lines of defence and passageways through which to defend their own territory and occasionally take action in neighbouring states (Sahlins 1990, p. 1433). The French Revolution has traditionally been seen as the benchmark that eventually boosted and gave further legitimacy to natural boundaries across Europe (Foucher 1991). In this context, the idea of natural boundaries not only served to lay the foundations for important diplomatic agreements with other kingdoms and neighbouring states, but also broadly defined the ideal territorial objectives of many states' foreign policy.

As has recently been proved (García-Álvarez and Puente-Lozano, 2015b, 2017a, 2017b), and as examined in Chaps. 6, 7, and 8, around the same time the Iberian monarchs also used natural boundaries in their border delineation practices. For example, the Boundary Treaties of Madrid (1750) and San Ildefonso (1777) indicated river courses and other natural features as the most suitable elements for serving as fixed and permanent limits when stipulating how the boundaries between Spanish and Portuguese possessions should be drawn up in colonial America (Zusman 2000; García-Álvarez and Puyo 2015). Similarly, throughout the nineteenth century, the virtues previously attributed to natural borders, or, for the purposes of tracing new internal administrative divisions, to natural boundaries, continued to be defended on an international level (Raztel 1897) and were very much in accordance with the tradition of the modern law of nations. They therefore figured prominently in international law and treaty practices around the globe (Bouchez 1963; Caflisch 1989; Teclaff 1991).

However, as Chaps. 6, 7, and 8 make clear, the more the accurate clarification of the natural boundary was sought and attempted, the more new and numerous unanticipated problems emerged as the social, historical, and geographical intricacies of the actual reality of these fluvial or mountainous spaces became evident. Interestingly, many recent works (Normand 1998; Desplat 2002; Catala et al. 2011; Salvatici 2005; Donaldson 2008; Kaplan et al. 2009; di Fiore and Meriggi 2013) have pointed out that tracing the boundary between two countries based on natural features was much less stable, more contested, and far more impractical and abstract than it appeared in theory or in maps (Sahlins 1990, p. 1441).

In this sense, it is worth noting that even though natural borders were to fulfil at that time the important role of marking out states' limits and their separation, borderlands were fully fledged connected transnational spaces, as local customs, inter-community agreements, and legal practices regulating access to resources and wider patterns of land possession had long given shape to these spaces, underlying traditional subsistence economies (Herzog 2015).

Again, boundary commissions are argued here to have been key actors in managing and coming to terms with the specific realities of these borderlands, which enabled pragmatic and highly context-sensitive arrangements to emerge. In this sense, it is worth pointing out that the general acceptance of natural border theory was already much more qualified in terms of practices and discourses concerning territory in the second half of the nineteenth century (at least in comparison with earlier historical periods). As a matter of fact, reports and discussions held throughout boundary-making processes, as well as the wider geographical, political, and technical scholarship of the time, are marked by an intense awareness of the zonal character of certain borderlands separated by mountains and watercourses—for example, the assumption that in geographical, economic, and social terms, such areas functioned in practice as a single unit rather than as a boundary line, since river courses and mountain valleys in modern societies were spaces propitious to settlement, exchange, and joint development by the inhabitants of both sides (Burgueño 1996, pp. 232–234; García-Álvarez 2002, pp. 264–267). This was actually heightened by the tracing and fixing of borders, not the other way around.

Against this theoretical backdrop, the chapters in the second part offer a telling example of the need to carefully analyse the geohistorical production of borders (particularly of 'natural' borders) by closely examining the material and representational practices that brought about territory on the ground. In Chap. 6, Perla Zusman and Sergio Nunes, by fully exploring the active role played by the boundary commissions responsible for delimiting the border between Spanish and Portuguese possessions in the Rio de la Plata region, in application of the 1750 Treaty of Madrid, provide an interesting example of how diplomatically agreed boundaries were 'resignified'. This was so because the actual geographical and social complexities of the borderlands (namely a highly complex hydrological system with significant current flow variations and sediment accumulation on the riverbed that entailed a branching of the mainstream channel), as well as issues with local populations, were highlighted by extensive fieldwork. Discussions on how to fix the boundary in the Ibicuy River and how to deal with the local inhabitants of these borderlands are analysed in this chapter, given their centrality in the process and how prominently they figured within the specific cartography produced by the commissioners.

In Chap. 7, Jean-Yves Puyo addresses boundary-making along the Franco–Spanish border by putting the spotlight on the two different rationales that have underlain boundary-making over the centuries: rights of use over land shaping the territorial pattern of borderlands vs. the very much praised theory of natural boundaries, a concept widely promoted in Europe by the French Revolution. Situated on a mountain massif, namely the Pyrenees, this boundary had but very recently been established (in the mid-nineteenth century), leading to the eventual separation of local communities that had been united for centuries in kingdoms straddling both sides of the mountain range. Although in some cases, as in the dispute over the Iraty massif discussed here, the drawing of the modern boundary involved the imposition of state interests, the chapter foregrounds the complex relationships woven between the representatives of the central governments and those of local communities, and how the territorial vision of local communities played an important role in the outcome of the final arbitrations made by the commissions.

In Chap. 8, Benjamin Duinat analyses discussions held throughout the 1890s between French and Spanish state representatives on the new layout of the 'dividing line' in the Bidasoa River that was to be fixed following the *thalweg* according to the Treaty of Bayonne (1856). The more the separation of territorial jurisdiction was refined, the more new unanticipated problems emerged, and eventually in 1893 the French Minister of the Navy noted that due to a recent storm the river regime had changed so much that it was necessary to conduct a new delimitation. Somehow, the sovereignty line was adrift. The chapter goes on to fully examine the role played by the numerous and different actors—for example, boundary commissioners, ministries, intermediary and local agents, border populations —and how both the lack of accurate delimitations within this fluid aquatic space and the urgent demand that the state provide a lasting delimitation were used by local populations to their advantage alternately and depending on their changing interests.

In Chap. 9, Luis Urteaga examines the fieldwork carried out by the Spanish section of the Joint Boundary Commission that attempted, unsuccessfully, to delimit the Franco–Spanish border in Morocco between 1913 and 1933. The history of this failure illustrates, in exemplary fashion, some of the difficulties typically associated with border-making processes in colonial Africa: the seemingly unequivocal lines drawn up in diplomatic treaties became uncertain boundaries in the landscape; expert commissions operated with less knowledge of the environment, with all of its complex topographical, linguistic, and cultural variations, and with far less regard for the needs and interests of the local society than delimitation commissions in Europe; there were no good maps of the area and there was no geodetic network to support topographical operations; and control of the territory (either by the main Moroccan authorities or by the two colonial powers) was precarious, with commissions exposed to hostility from local rebel tribes. In addition to this, the shifting colonial strategies brought about by a changing geopolitical context increased the disagreements between the Spanish and French governments and prevented the ratification and completion of the Commission's work. Paradoxically, although the maps drawn up by the Boundary Commission of Morocco were mostly consigned to the archives and ignored by Spanish historiography, one of them, carried out by Lieutenant Colonel Federico Montaner Canet in 1932, which summarized the border lines supposedly agreed between the French and Spanish governments from 1902 onwards, implicitly blamed France for the progressive reduction of the territory assigned to the Spanish Protectorate in Morocco, and in the following decades became 'fuel for nationalist rhetoric' in Spain.

Finally, in the last essay of this book (Chap. 10) Laura di Fiore provides some concluding remarks by explicitly framing concerns about borders and border-making within the field of global history, and thus further elaborating on theoretical issues (some of them pointed at the beginning of this chapter) that have come to the forefront in recent debates raised in the context of the spatial turn. Di Fiore argues that global history's sensitivity towards the category of space and its sound attempt to go beyond methodological nationalism enable a new twofold reading of the boundary-making processes having taken place throughout eighteenth and nineteenth centuries. On the one hand, global history allows for detaching boundary-making processes that developed in Europe at that time from hegemonic state-centred narratives about the rise of the nation-state, which conventionally entail a top-down perspective. Instead, and in the light of the premises of the 'spatial turn', eighteenth and nineteenth centuries boundary-making can be envisioned as a moment of production of a new space, in which various social actors besides the state took part. On the other hand, global history's interest in regional and transregional spaces may lead to a fruitful dialogue with the concept of region and the ethnographic methods developed by anthropology and geography within border studies over the last years, in order to develop new perspectives on borderlands. In that sense, this enhanced awareness of the specific production and reproduction of space that occurs in border-making processes and through the very dynamics of borderlands envisioned as complex transnational spaces sheds new light on the modern history of European and colonial territoriality. Therefore, the concluding chapter intends to flesh out the type of theoretically informed and simultaneously empirically grounded perspective that enables to geohistoricize modern cultures and practices of boundary-making, as this books has undertaken to do.

References

Adelman A, Aron S (1999) From borderlands to borders: empires, nation-states, and the peoples in between in North American history. Am Hist Rev 104(3):814–841

African Union Border Programme (2014) Delimitation and demarcation of boundaries in Africa. General issues and case studies. Commission of the African Union, Department of Peace and Security, Addis Ababa

Agnew J (1994) The territorial trap: the geographical assumptions of international relations theory. Rev Int Polit Econ 1(1):53–80

Agnew J (2003) Geopolitics: re-visioning world politics. Routledge, London

Agnew J (2005) Sovereignty regimes: territoriality and state authority in contemporary world politics. Ann Assoc Am Geogr 95(2):437–461

Agnew J (2009) Globalization and sovereignty. Rowman & Littlefield Publishers, Lanham, Md

Akerman J (1995) The structuring of political territory in early printed atlases. Imago Mundi 47:138–154

Alcalde A (2018) Spatializing transnational history: European spaces and territories. European Review of History: Revue Européenne d'histoire 25(3–4):553–567

Amelina A, Nergiz DD, Faist Th, Glick Schill N (2012) Beyond methodological nationalism: research methodologies for cross-border studies. Routledge, London

Ancel J (1938) Géographie des frontières. Gallimard, Paris

Antunes C, Polónia A (eds) (2016) Beyond Empires: global, self-organizing, cross-imperial networks, 1500–1800. Brill Academic Publishing, Leiden

Aron S (2006) American confluence: The Missouri frontier from Borderland to border state. Indiana University Press, Bloomington

Arvizu F (1992) El conflicto de los Alduides (Pirineo Navarro). Estudio institucional de los problemas de límites, pastos y facerías según la documentación inédita de los archivos franceses (siglos XVII-XIX). Gobierno de Navarra, Departamento de Presidencia, Pamplona

Arvizu F (1997). La solución institucional del conflicto de los Alduides (Pirineo Navarro). Estudio de las negociaciones entre Francia y España desde 1829 a la conclusión del Tratado de Límites de 1856, según la documentación inédita de los Archivos de los Ministerios de Asuntos Exteriores de Francia y España. Universidad de León, León

Arvizu F (2001) Frontera y fronterizos: el caso de los Alduides (Pirineo de Navarra). Anuario de Historia del Derecho Español LXXI, pp 9–48

Ates S (2013). Ottoman-Iranian borderlands. Making a boundary 1843–1914. Cambridge University Press, Cambridge

Axtmann R (2004) The state of the state: the model of the modern state and its contemporary transformation. Int Polit Sci Rev/revue Internationale De Science Politique 25(3):259–279

Barret T (1999) At the edge of empire: The Terek Cossacks and the north Caucasus frontier, 1700–1860. Westview Press, Boulder

Bartleson J (1995) A genealogy of *sovereignty*. Cambridge University Press, Cambridge

Bartov O, Weitz ED (2013) Shatterzones of empire. Coexistence and violence in the German, Habsburg, Russian, and Ottoman borderlands. Indiana University Press, Bloomington

Baycroft T (2004) Culture, identity and nationalism: french flanders in the nineteenth and twentieth centuries. The Boydell Press, New York

Benton L, Ross RJ (eds) (2013) Legal pluralism and empires, 1500–1850. New York University Press, New York

Benton L (2010) A search for sovereignty. Law and geography in European empires, 1400–1900. Cambridge University Press, Cambridge

Bernat R, Callafat G (2018) La microhistoire globale: affaire(s) à suivre. Annales 73(1):3–18

Biggs M (1999) Putting the state on the map: cartography, territory, and European state formation. Comp Stud Soc Hist 41(2):374–405

Bjork JE (2008) Neither German nor Pole: Catholicism and national indifference in a central European borderland. University of Michigan Press, Ann Arbor

Black J (1997) Maps and history: constructing images of the past. Yale University Press, New Haven

Blackhawk N (2006) Violence over the Land. Indians and empires in the early American West. Harvard University Press, Cambridge

Blais H (2011) An intra-imperial conflict: the mapping of the border between Algeria and Tunisia, 1881–1914. J Hist Geogr 37(2):178–190

Blais H (2014) La longue histoire de la délimitation des frontières de l'Algérie. In: Bouchène A, Peyroulou JP, Tengour OS, Thénault S (eds) Histoire de l'Algérie à la période coloniale, 1830–1962. La Découverte, pp 100–113

Blomley N (2003) Law, property and the spaces of violence: the frontier, the survey, and the grid. Ann Assoc Am Geogr 93(1):121–141

Bouchez LJ (1963) The fixing of boundaries in international boundary rivers. Int Comp Law Q 12:789–817

Branch J (2011) Mapping the Sovereign state: technology, authority, and systemic change. Int Organ 65(1):1–36

Branch J (2013) The cartographic state. Cambridge University Press, Cambridge

Branch J (2017) Territory as an institution: spatial ideas, practices and technologies. Territory Politics Governance 5(2):1–14

Breyfogle N (2005) Heretics and colonizers: forging Russia's empire in the South Caucasus. Cornell University Press, Ithaca

Brigham AP (1919) Principles in the determination of boundaries. Geogr Rev 7(4):201–219

Brighenti A (2006) On territory as relationship and law as territory. Can J Law Soc 21(2):65–86

Brooks JF (2002) Captives and cousins: slavery, kinship, and community in the southwest borderlands. Omohundro Institute of Early American History and Culture and the University of North Carolina Press, Williamsburg, Virginia
Brunhe J, Vallaux C (1921) La géographie de l'historie. Félix Alcan, Paris
Burbank J, Cooper F (2011) Empires in world history: power and the politics of difference. Princeton University Press, Princeton
Burgueño J (1996) Geografía política de la España constitucional. La división provincial. Centro de Estudios Constitucionales, Madrid
Byrnes B (2001) Boundary markers: land surveying and the colonisation of New Zealand. Bridget Williams Books Limited, Wellington
Caflisch L (1989) Regles generales du droit des curs d'eau internationaux. Recueil Des Curs De L'academie De Droit International De La Haye 219(7):9–224
Campbell CJ (2018) Space, place and scale: human geography and spatial history. Past Present 239(1):23–45
Capdevila J (2009) Historia del deslinde de la frontera hispano-francesa. Del tratado de los Pirineos (1659) a los tratados de Bayona (1856–1868). Centro Nacional de Información Geográfica, Madrid
Carrol FM (2001) A good and wise measure: the search for the Canadian-American boundary, 1783–1842. University of Toronto Press, Toronto
Carter D (2007) State visions, landscape, and disease: discovering malaria in Argentina, 1890–1920. Geoforum 39(1):278–293
Catalá M, Le Page D, Meuret JC (2011) Frontières oubliées, frontières retrouvées. Presses Universitaires de Rennes, Rennes, Marches et limites anciennes en France et en Europe
Chavarría F (2005) Justicia y estrategia: teoría y práctica de las leyes de la guerra en un contexto fronterizo. Melanges De La Casa De Velazquez 35:185–215
Conrad S (2016) What is global history? Princeton University Press, Princeton
Craib RB (2004) Cartographic Mexico: a history of state fixations and fugitive landscapes. Duke University Press, Durham
Crampton J (2010) Cartographic calculations of territory. Prog Hum Geogr 34(1):92–103
De la Teja JF, Frank R (2005) Choice, persuasion and coercion. Social control on Spain's North American frontiers. University of New Mexico Press, Albuquerque
De Vito CG (2019) History without scale: the micro-spatial perspective. Past & Present 242(14):348–372
De Vries J (2019) Playing with scales: the global and the micro, the macro and the nano. Past & Present 14:23–36
Demandt A (1990) Deutschlands Grenzenin der Geschichte. Deck, Munich
Desplat Ch (2002) Frontières. CTHS, París
Di Fiore L, Mariggi M (2013) World history. Le nuove rotte della storia. Laterza, Bari
Di Fiore L (2015) Alla frontiera. Confini e documenti di identità nel Mezzogiorno continentale preunitario. Rubbetino Editore, Soveria Mannelli
Di Fiore L (2017) The production of borders in nineteenth-century Europe: between institutional boundaries and transnational practices of space. European Review of History: Revue Européenne D'histoire 24(1):36–57
Dias MH (2009) Finis Portugalliae - Nos confins de Portugal. Cartografía militar e identidad territorial. Instituto Xeográfico do Exército, Lisbon
Diener AC, Hagen J (2009) Theorizing borders in a 'Borderless World': globalization, territory and identity. Geogr Compass 3(3):1196–1216
Donaldson JW (ed) (2008) Politics and scale in boundary-making: the work of boundary commissions (special issue). J Hist Geogr 34:393–515. https://www.sciencedirect.com/journal/journal-of-historical-geography/vol/34/issue/3
Döring J, Thielmann T (eds) (2008) Spatial Turn: das Raumparadigma in den Kultur- und Sozialwissenschaften. Transcript, Bielefeld
Dornel L (2002) La frontière (le voisin) et l'étranger. Les enjeux identitaires d'un conflit frontalier. Revue d'histoire du XIXe siècle 1 24:11–24
Duinat B (2019) Une anomalie frontalière à la loupe : Ondarrola, un hameau entre la France et l'Espagne (v. 1780 – v. 1860). Revue D'histoire Du XIXe Siècle 59(1):249–268
Dunlop CT (2015) Cartophilia: maps and the search for identity in the French-German borderland. University of Chicago Press, Chicago
Elden, S (2005) Missing the point about globalization. Globalization, deterritorialization and the space of the world. Trans Inst Br Geogr 30(1):8–19
Elden S (2006) Speaking against number: Heidegger, language and the politics of calculation. Edinburgh University Press, Edinburgh
Elden S (2007a) Governmentality, calculation, territory. Environ Plan D Soc Space 25(3):562–580
Elden S (2007b) Rethinking governmentality. Polit Geogr 26(1):29–33
Elden S (2010) Thinking territory historically. Geopolitics 14(4):757–761
Elden S (2013a) The Birth of Territory. The University of Chicago Press, Chicago
Elden S (2013b) How should we do the history of territory? Territory Politics Governance 1(1):5–20
Evans RJW (2006) Frontiers and National identities in Central European-history. In: Evans RJW (ed) Austria, Hungary and the Habsburgs. Central Europe c.1683–1867. Offord university Press, Oxford, pp 114–133
Faist Th (2012) Toward a transnational methodology: methods to address methodological nationalism, essentialism, and positionality. Rev Eur Migr Int 28 (1):51–70
Fall J (2005) Drawing the line. Nature, hybridity, and politics in transboundary spaces. Ashgate, Aldershot
Fall J (2010) Artificial states? On the enduring geographical myth of natural borders. Polit Geogr 29(3):140–147
Ford L (2010) Settler sovereignty. Jurisdiction and indigenous people in America and Australia, 1788–1836. Harvard University Press, Cambridge

Foucault M (1978) Right of death and power over life. In: Foucault M The history of sexuality, vol I. Translated from the French by Robert Hurley. Pantheon Books, New York, pp 133–160

Foucher M (1991) Fronts et frontières. Un tour du monde géopolitique. Fayard, Paris

Fratini M (2003) Storie di frontiera e pratiche di confine nelle valli valdesi. La Beidana 47:1–18

Freist D (2017) A global microhistory of the early modern periods: social sites and the interconnectedness of human lives. Quaderni Storici 155, LII(2):537–555

Galmen A (2008) The emigration state and the modern geopolitical imagination. Polit Geogr 27(8):840–856

García J (1996) A configuracão da fronteira luso-espanhola nos mapas dos séculos XV a XVIII. Treballs de la Societat Catalana de Geografia 11 (41):1133–2190

García-Álvarez J (2002) Provincias, regiones y comunidades autónomas: la formación del mapa político de España. Secretaría General del Senado, Madrid

García-Álvarez J (2006) Geografía regional. In: Hiernaux D, Lindón A (eds) Tratado de Geografía Humana. Anthropos, Barcelona, pp 25–70

García-Álvarez J (2019) Geopolíticas de la territorialidad moderna: el Tratado hispano-portugués de Límites de 1864 y la extinción del Couto Mixto (1840–1868). Boletín de la Asociación de Geógrafos Españoles (82). https://doi.org/10.21138/bage.2798

García-Álvarez J, Puente Lozano P (2015a) Las Comisiones Mixtas de Límites y las representaciones geográficas de la frontera hispano-portuguesa (1855–1906). Revista de Historiografía 23:67–100. https://e-revistas.uc3m.es/index.php/REVHISTO/article/view/2756

García-Álvarez J, Puente-Lozano P (2015b) La delimitación de la frontera luso-española en el río Miño (1855–1906). Conflictos territoriales y representaciones cartográficas", Terra Brasilis (Nova Série) 6. https://doi.org/10.4000/terrabrasilis.1708

García-Álvarez J, Puente-Lozano P (2017a) A controversial natural border. The making of the Spanish-Portuguese boundary along the Minho River (1855–1906). Geoforum 79:5–16

García-Álvarez J, Puente-Lozano P (2017b) Bridging central state and local communities' territorial visions: boundary commissions and the making of Iberian borders, 1750–1900. J Hist Geogr 56(2):345–367

García-Álvarez J, Puyo JY (2015) La aportación geográfica y cartográfica de las Comisiones de Límites luso-franco-españolas (siglos XVIII y XIX). Terra Brasilis (Nova Série) [online] 6. https://doi.org/10.4000/terrabrasilis.1693

García-Álvarez J, Puyo J-Y (2019) Los primeros intentos de delimitación moderna de la frontera pirenaica: la Comisión franco-española Caro-Ornano y su legado cartográfico (1984–1792). Revista de Historiografia 30:15–44

Glacken C (1967) Traces on the Rhodian shore: nature and culture in western thought from ancient times to the end of the eighteenth century. University of California Press, California

Godinho P (2011) Oír o galo cantar dúas veces. Deputación Provincial, Ourense

Goettlich K (2019a) The rise of linear borders in world politics. Eur J Int Rel 25(1):203–228

Goettlich K (2019b) *From frontiers to borders: the origins and* consequences of linear *borders* in international politics. Dissertation, The London School of Economics and Political Science (LSE), London

Guy NC (2008) Linguistic boundaries and geopolitical interests: the Albanian boundary commissions, 1878–1926. J Hist Geogr 34(1):448–470

Guy DJ, Sheridan TE (1998) Contested ground. Comparatives frontiers on the northern and southern edges of the Spanish Empire. University of Arizona Press, Tucson

Hämäläinen P, Johnson B (2012) Major problems in the history of North American borderlands. Cengage Learning, Boston

Hannah M (2000) Governmentality and the mastery of territory in nineteenth-century. Cambridge University Press, Cambridge

Hannah M (2009) Calculable territory and the west German census boycott movements of the 1980s. Polit Geogr 28(1):66–75

Hekster O, Kaizer T (2009) Frontiers in the Roman World. Brill, Leiden and Boston

Herzog T (2015) Frontiers of possession: Spain and Portugal in Europe and the Americas. Cambridge University Press, Cambridge

Hobson J (2009) The eurocentric conception of world politics: western international theory, 1760–2010. Cambridge University Press, Cambridge

Jerram L (2013) Space: a useless category for historical analysis? History Theory 52(3):400–419

Jesoop B (2007) The state and state power. In: Clegg S, Haugaard M (eds) The SAGE handbook of power. SAGE, London, pp 367–382

Johnson B, Graybill AR (2010) Bridging National borders in North America: transnational and comparative histories. Duke University Press, Durham

Kanska KA, Manko RT (2003) Shifts in international river boundaries. Polish Yearbook Int Law 11(26): 135–155

Kaplan B, Carlson, M, Cruz L (eds.) (2009) Boundaries and their meanings in the history of the Netherlands. Brill, Boston-Leiden

Keating J (2018) Invisible *countries*: Journeys to the edge of nationhood. Yale University Press, New Haven and London

Krassner SD (1999) Sovereignty: organized hypocrisy. Princeton University Press, Princeton

Kulczycki JJ (2016) Belonging to the nation: inclusion and exclusion in the Polish-German borderlands, 1939–1951. Harvard University Press, Cambridge

Laboulais I (2008) Les usages des *cartes* (XVIIe-XIXe siècle). Pour une approche pragmatique des productions cartographiques. Press universitaires de Strasbourg, Strasbourg

Langfur H (2006) The forbidden lands. Colonial identity, frontier violence, and the persistence of Brazil's Eastern Indians, 1750–1830. Stanford University Press, Stanford
Lefebvre C (2011) We have tailored Africa: French colonialism and the "artificiality" of Africa's *borders* in the interwar period. J Hist Geogr 37(2):191–202
Lewis MW, Wigen K (1997) The myth of continents: a critique of metageography. University of California Press, Berkeley
Lois C (2019) De las fronteras coloniales del imperio hispánico en América a los límites internacionales entre Estados latinoamericanos independientes: génesis de la imposibilidad de un mapa político de Sudamérica consensuado. Revista De Historiografía 30:207–222
Lossau J, Lippuner R (2004) Geographie und Spatial turn. Dispo 1 58(3):201–211
Maier Ch (2016) Once within borders. Territories of power, wealth, and belonging since 1500. The Belnak Press, New York
Mandrini R, and Paz C D (2003) (eds.) Las fronteras hispanocriollas del mundo indígena latinoamericano en los siglos XVIII-XIX: Un estudio comparativo. CEHIR/IEHS/UNS, Tandil.
Mandrini R (2006) Vivir entre dos mundos. Las fronteras del Sur del a Argentina. Siglos XVII y XIX. Taurus, Buenos Aires
Meriggi M (2016) Racconti di confine: Nel Mezzogiorno del Settecento. Il Mulino, Bolonia
Middell M, Naumann K (2010) Global history and the spatial turn: from the impact of area studies to the study of critical junctures of globalization. J Glob Hist 5(01):149–170
Minghi J (1963) Boundary studies in political geography. Ann Assoc Am Geogr 53(3):407–428
Misra S (2011) Becoming a borderland: the politics of space and identity in colonial Northeastern India. Routledge India, New Delhi
Mongiano E (1991) La delimitazione dei confini dello Stato: attività diplomatica *e* produzione cartografica nei territori sabaudi (1713–1798). Studi Piemontesi XX 1:45–56
Morieux R (2016) The channel: England, France and the construction of a maritime border in the eighteenth century. Cambridge University, Cambridge
Müller M, Torp C (2009) Conceptualising transnational spaces in history. European Review-Revue Européenne D'histoire 16(5):609–617
Murdock CE (2011) Changing places: society, culture, and territory in the Saxon-Bohemian borderlands, 1870–1946. University of Michigan Press, Ann Arbor
Nordman D (1998) Frontières de France: De l'espace au territoire, XVIe–XIXe siècle. By Daniel Nordman. Bibliothèque des Histoires. Gallimard, Paris
Nuñez A, Zusman P, Hevilla MC (2017) La producción social de las fronteras (Des-naturalizando las fronteras políticas en América Latina) Rev. geogr. Norte Gd. 66. https://doi.org/10.4067/S0718-34022017000100001
O'Brien P (2006) Historiographical traditions and modern imperatives for the restoration of global history. J Glob Hist 1(1):3–39
Osiander A (2001) Sovereignty, international relations, and the Westphalian *Myth*. Int Organ 55(2):251–287
Paasi A (2011) Border theory: an unattainable dream or a realistic aim for border scholars? In: Wasti-Walter D (ed) The Ashgate research companion to border studies. Ashgate, London, pp 11–31
Paasi A (2020) From bounded spaces to relational social constructs: conceptualization of the region in geography. In: Kohlenberg P, Godehardt N (eds) The multidimensionality of regions in world politics. Routledge, London, pp 1–19
Painter J (2010) Rethinking territory. Antipode 42(5): 1090–1118
Parker N, Vaughan-Williams N (2013) Critical border studies: broadening and deepening the 'lines in the sand' agenda. Routledge, London
Patel KK (2015) An *Emperor* without clothes? The Debate about *transnational* history twenty-five years on. Histoire@Politique, 26. Retrieved from: https://www.histoire-politique.fr/documents/26/pistes/pdf/HP26-Pistesetdebats_Kiran_Patel_def.pdf. Accessed Nov 2020
Philipott D (2001) Revolutions in *sovereignty*: how ideas shaped modern international relations. Princeton University Press, Princeton
Polak-Springer P (2015) Recovered territory: a German-Polish conflict over land and culture, 1919–1989. Berghahn, New York
Poujade P (1998) Les relacions transpirinenques a la Catalunya Moderna: els tractats de "Ligues i Patzeries. Pedralbes: Revista D'historia Moderna 18(1):189–200
Prescott V, Triggs GD (2008) International frontiers and boundaries: law, politics and geography. Martinus Nijhoff Publishers, Leiden and Boston
Puyo J-Y, García-Álvarez J (2016) Une frontière pour les Pyrénées: l'épisode trop méconnu de la Comission topographique franco-espagnole Caro-Ornano (1784–1792). Cartes et Géomatique 228:47–58
Rankin KJ (2008) The role of the Irish boundary commission in the entrenchment of the Irish border: from tactical panacea to political liability. J Hist Geogr 34(3):422–447
Rankin KJ, Schofield R (2004) The troubled historiography of classical boundary terminology. Institute for British-Irish Studies Working paper, 41
Raviola BA (2010) De l'osmose à la séparation. La construction de la frontière entre la France et le Piémont-Savoie (XVIe- XVIII^{e}siècles). Cahiers De La Méditerranée 81:271–289
Raztel F (1897) Politische Geographie. Oldenburgh, Munich
Readman P, Radding C, Bryant C (2014). Borderlands in World History, 1700–1914. Palgrave, Chennai
Rebert P (2001) La gran línea. Mapping the US-Mexico boundary. University of Texas Press, Austin
Revel J (1996) Jeux d'echelles. La micro-analyse à l'expérience. Gallimard et Le Seuil, Paris

Ricuperati G (2001) Lo stato sabaudo nel Settecento. Dal trionfo delle burocrazie alla crisi d'antico regime. UTET, Lilan

Rieber AJ (2014) The struggle for the Eurasian borderlands. From the rise of early modern empires to the end of the First World War. University of Cambridge Press, Cambridge

Rose-Redwood RS (2012) With numbers in place: security, territory, and the production of calculable space. Ann Assoc Am Geogr 102(2):295–319

Rowe M (2014) Borders, war, and nation-building in Napoleon's Europe. In: Readman P, Radding C, Bryant C (eds) (2014) Borderlands in World History, 1700–1914. Palgrave, Chennai, pp 143–165

Ruggie J (1993) Territoriality and beyond: problematizing modernity in international relations. Int Organ 47 (1):139–174

Sachsenmaier D (2012) Global perspectives on global history. Theories and approaches in a connected world. Cambridge University Press, Cambridge

Sahlins P (1989) Boundaries. The making of France and Spain in the Pyrenees. University of California Press, Berkeley and Los Angeles

Sahlins P (1990) Natural frontiers revisited: France's boundaries since the seventeenth century. Am Hist Rev 95:1423–1451

Sahlins P (1998) State formation and national identity in the Catalan borderlands during the eighteenth and nineteenth centuries. In: Wilson TM, Donnan H (eds) Border identities: Nation and state at international frontiers. Cambridge University Press, Cambridge, pp 31–61

Salvaticci S (ed) (2005) *Confini*. Costruzioni, attraversamenti, rappresentazioni. Rubbettino Editores, Soveria Mannelli

Sandouno M (2014) Une histoire des frontières guinéennes (années 1880-2010): héritage colonial, négociation et conflictualité. Dissertation, Université Toulouse le Mirail, Toulouse

Sassen S (2013) When territory deborders territoriality. Territory Politics Governance 1(1):21–45

Schaffter M, Fall J, Debarbieux B (2010) Unbounded boundary studies and collapsed categories: rethinking spatial objects. Prog Human Geogr 34(2):254–262

Schendel W (2005) The Bengal borderland. Beyond state and nation in South Asia. Anthem Press, London

Schoefield R (2015) Back to the barrier function: where next for international boundary and territorial disputes in political geography? Geography 100(3):133–143

Schroeter F (2007) Les frontières de la Suisse: Questions choisies. Schullthess, Zürich

Schulz CA (2019) Territorial sovereignty and the end of inter-cultural diplomacy along the "southern frontier." Eur J Int Rel 25(3):878–903

Sermet J (1983) La Frontière hispano-française des Pyrénées: et les conditions de sa délimitation. Marrimpouey, Pau

Shah N (2011) The territorial trap of the territorial trap: global transformation and the problem of the state's two territories. Int Political Sociol 6(1):57–76

Sheehan JJ (2006) The *problem* of sovereignty in European history. Am Hist Rev 111(1):1–15

Sletto B (2001) Conservation planning, boundary-making and border terrains: the desire for forest and order in the Gran Sabana, Venezuela. Geoforum 42(2):197–210

Steinberg PE (2005) Insularity, sovereignty and statehood: the representation of islands on Portolan Charts and the construction of the territorial state. Geografiska Annaler. Series B, Human Geogr 87(4):253–265

Stoklosa K, Besier G (2014) European border regions in comparison. Overcoming nationalistic aspects or renationalization? Taylor & Francis, New York

Strandsbjerg J (2008) The cartographic production of territorial space: mapping and state formation in early modern Denmark. Geopolitics 13(2):335–358

Strandsbjerg J (2010) Territory, globalization and international relations. The cartographic reality of space. Palgrave, London

Strandsbjerg J (2012) Cartopolitics, geopolitics and boundaries in the Arctic. Geopolitics 17(4):818–842

Teclaff LA (1991) Fiat on custom: the checkered development of international water law. Nat Resour J 31 (1):46–73

Teschke B (2003) The myth of 1648: class, geopolitics and the making of modern international relations. Verso, London

Tilly Ch (1992) Coercion, capital, and European states. Blackwell Publishers, Cambridge

Turco A (2019) Micro/macro: ¿local/global? El problema de la localidad en una historia espacializada. Historia Crítica 69:37–67

Warf B, Arias S (eds) (2008) The spatial turn: interdisciplinary perspectives. Routledge, New York

Weber DJ (1992) The Spanish frontier in North America. Yale University Press, Boston

Werner M, Zimmermann B (2006) Beyond comparison: *histoire croisée* and the challenge of reflexivity. History Theory 45(1):30–50

Zeller G (1933) La Monarchie de l'ancienne regime et les frontieres naturelles. Revue d'Histoire Moderne VIII:305-333

Zusman P (2006) Geografías Históricas y fronteras. In: Hiernaux D, Lindón A (eds) Tratado de Geografía Humana. Anthropos, Barcelona, pp 170–186

Zusman P, Barros C (2000) Nuevas y viejas fronteras. ¿Nuevos y viejos encuentros y desencuentros? Scripta Nova. Revista Electrónica de Geografía y Ciencias Sociales. 69(50). Retrieved from http://www.ub.edu/geocrit/sn-69-50.htm

Part I

Boundary Treaties and Commissions: Theoretical Foundations and Praxis on the Ground

2 Modern State Sovereignty as Geometric Purification of Territory: Delimitation Theories and Practices in the 1864 Spanish–Portuguese Boundary Treaty

Jacobo García-Álvarez and Paloma Puente-Lozano

Abstract

The 1864 Spanish–Portuguese Boundary Treaty provides an exemplary reflection of the concept of sovereignty characteristic of liberal states, and more broadly, of the consolidation of the territorial state in the Iberian Peninsula. As opposed to the concept of sovereignty still in force under absolute monarchies, which was dynastic, patrimonial, and jurisdictional, the liberal governments responsible for the Treaty based state sovereignty on the exclusive and indivisible jurisdiction and control of territory, which linear boundaries would precisely delimit. Drawing on an analysis of mostly unpublished archival documentation preparatory to the 1864 Treaty, this chapter consists of three sections. Firstly, it addresses the concept and genesis of the territorial state, as well as the ideas of sovereignty and borders linked to it, in Europe and the Iberian countries. Secondly, it analyzes the theoretical and general delimitation criteria that guided the Treaty. Finally, it studies the way in which these criteria were applied in practice for resolution of the various territorial issues addressed in the Treaty. The main thesis of this work is that the 1864 Treaty, despite not strictly creating a linear boundary on the Spanish–Portuguese border, did give it an international legal character, in addition to completing and clarifying the border to delimit state territory in a continuous and compact way. Implementation of this process came by means of a series of operations consisting of a geometric and geopolitical purification of the territory.

This work has been supported by the Spanish Ministry of Economy, Industry, and Competitiveness and the European Regional Development Fund (Research Project CSO2015-65301-P, MINECO/FEDER), as well as by the Spanish State Research Agency (Research Project PID2020-114088GB-I00/AEI/10.13039/501100011033).

J. García-Álvarez (✉) · P. Puente-Lozano
Department of Humanities: History, Geography and Art, Carlos III University of Madrid, Getafe, Madrid, Spain
e-mail: jacobo.garcia@uc3m.es

P. Puente-Lozano
e-mail: ppuente@hum.uc3m.es

2.1 Introduction

Even though the territorial outline of the Spanish–Portuguese border—popularly known as *la Raya* in Spanish (or *a Raia* in Portuguese and Galician)—dates for the most part from medieval times, its exact, modern, and definitive delimitation was not legally established until the *Tratado de Límites* (Boundary Treaty, henceforth BT) of 1864 and the

J. García-Álvarez and P. Puente-Lozano (eds.), *Beneath the Lines*, Historical Geography and Geosciences, https://doi.org/10.1007/978-3-030-96904-2_2

Convenio de Límites (Convention of Limits) of 1926. The former, signed in Lisbon on September 29, 1864, establishes the border between the mouth of the Minho River and the confluence of the Caia and Guadiana rivers. The latter, also signed in Lisbon on June 29, 1926, delimits the border from the confluence of the Cuncos and Guadiana rivers to the latter's mouth in the Atlantic Ocean. Both treaties only left out the border defined by the Guadiana between the Caia and Cuncos rivers, which largely coincided with that of the municipality of Olivenza (Olivença in Portuguese). Although the latter came under Spanish sovereignty by virtue of the War of the Oranges and the treaties of Badajoz and Madrid of 1801, it has since been the subject of a lengthy diplomatic dispute between the two countries, which has been frozen and diluted in practice since they joined the European Economic Community on January 1, 1986.

The 1864 BT was approved during the reigns of Isabella II of Spain and Luís I of Portugal and was ratified by both countries' monarchs and their courts in 1865 and 1866, respectively. Its text consists of a preamble and thirty-one articles, and it was supplemented in 1866 by two annexes that, respectively, regulated boundary rivers and the seizure of livestock (United Nations 1991, pp. 275–289). Operations to execute the demarcation provided for in the BT went on for nearly forty years and came to a close with the *Acta General de Demarcación* (General Act of Demarcation), approved in Lisbon on December 1, 1906. As with the Boundary Treaties of Bayonne, signed between Spain and France from 1856 to 1868, which definitively delimited the border between the two countries, the 1864 BT expressed the firm will of the liberal states of Spain and Portugal to precisely, exclusively, and indivisibly delimit the territories under their sovereignty. In this regard, the BT attempted to eliminate a number of territorial disputes in areas where the border was not clearly delimited. As such, it put an end to the irregular situation of certain settlements or places through which the border ran (the so-called promiscuous villages), and it also entailed the disappearance of the Couto Mixto, a territory with an ambiguous political and administrative status, inherited from the Ancien Régime, which enjoyed numerous privileges and in many respects escaped the sovereignty of both kingdoms.

The main tasks regarding the preparation and subsequent execution of the1864 BT were undertaken by the Joint Boundary Commissions (henceforth JBCs), created between 1854 and 1855 and mainly made up of diplomats and military officers from both countries. These commissions were therefore entrusted with identifying, negotiating, drawing, delimiting, and representing with scientific precision a line that was not only a faithful reflection of the development of a modern notion of sovereignty and territory, but that would also enable such development.

The aim of this chapter is precisely to examine how, through its theoretical–practical balance, the 1864 BT understood and sought to materialize the connection between territory, sovereignty, and borders. More specifically, we set out to study the relationship between the ideas that emerged around these concepts in said process (from negotiating the instructions to delimit the border to the BT's final diplomatic discussions) and the very practices by means of which delimitation was implemented. According to our thesis, while the 1864 BT did not create a linear border in the Spanish–Portuguese *Raya*, it did provide it with an international legal status and, above all, completed and specified it, enabling the continuous and discrete delimitation of the territorial state through a series of operations that in a certain sense implied a geometric refinement or purification of this territory. Ultimately, this purification was key to the development and consolidation, in both Iberian countries, of a model of sovereignty and political organization of space that to this day remains dominant on a global scale, and which scholars have, significantly, termed the 'territorial state'.

This chapter is divided into three sections, in addition to this introduction and the conclusion. Firstly, it examines the concept and origins of territorial states in Europe and the Iberian countries, as well as the ideas of sovereignty and borders associated with them. Secondly, it analyzes the theoretical and general delimitation criteria that guided the 1864 BT. Finally, it studies the way in which said criteria were applied in practice to resolve the various territorial issues addressed by the BT. Apart from the specialist literature on this issue and some printed sources of the time, the research that has given rise to the present study is essentially based on the analysis of preparatory documentation for the 1864 BT, preserved in the National Historical Archive of Spain in Madrid (AHN), the vast majority of which remains unpublished, as well as in the Historical-Diplomatic Archive of the Portuguese Ministry of Foreign Affairs in Lisbon (AHDMNE), which had in large part already been published (Barreiros 1961–1965).

2.2 Territorial States, Borders, and Sovereignty in Europe and the Iberian Countries

Scholars dealing with the history of state formation and with the modern notion of territoriality in Europe have identified various stages and models that have characterized the evolution of the state, and of borders, from the Middle Ages to the present day (Storey 2001; Teschke 2006; Popescu 2012; Flint and Taylor 2018). In general, authors tend to distinguish between, on the one hand, jurisdictional and personal forms of rule and political authority, characteristic of the Middle Ages, and, on the other hand, sovereign statehood increasingly articulated upon territorial exclusivity. This last form, often referred to as the 'territorial state', having emerged in Western Europe in the sixteenth century, would go on to become the dominant state form in today's world, even though it would take several centuries to establish itself and spread, both within Europe and throughout the rest of the continents. As far as the jurisdictional forms of rule were concerned, they were typical of feudal society, based on personal relations between lords and vassals and the sovereign's control over people rather than over a territory. In medieval monarchies and political units, power could be shared between multiple centers of authority depending on the functions concerned (religious, fiscal, military, etc.). Sovereigns did not hold a monopoly on violence and conceived of the state as a personal asset that they could split up or expand through wars, inheritances, and marriages, resulting in unstable, discontinuous borders, often of a zonal rather than linear nature. Conversely, the territorial state would be characterized by the existence of a sovereign power exercised exclusively over a discrete and continuous territory, delimited with precision by linear borders.

The diplomatic origins of the territorial state are usually traced back to the Peace of Westphalia (1648), which ended the Thirty Years' War. With the aim of guaranteeing order, security, and stability, the states that were parties to the treaties associated with said peace agreed that each monarch or ruler would thereafter be recognized as the principal or exclusive sovereign over their territory (as opposed to imperial or papal authority, for instance). This meant that interference in a country's internal affairs would be deemed contrary to international law. However, the state system derived from Westphalia retained some typical features of the feudal period, such as the fact that states continued to be considered the hereditary property of monarchs (Teschke 2006).

Indeed, this hereditary conception would persist until the emergence of nationalisms and the principle of national sovereignty as proclaimed by the French Revolution, which transferred the power of the state from the monarch to the entire nation, understood as the body of citizens belonging to the territory of the state. It thus established a new concept of the state, namely the *nation-state*, which some authors consider to be a mere variant or evolution of the territorial state. By virtue of this concept, the territory of the state is subsumed into that of the nation, and citizens are required to acquire a nationality based largely (albeit not always, or not solely) on their place of birth or their territorial residence. In Europe, the shift toward a territorial conception of sovereignty and a linear vision of borders would be enshrined, on the diplomatic level, in the Congress of Vienna and the Treaties of Paris of 1815 to 1816. This conception translated into the establishment of modern boundary treaties, which implied suppressing both extant non-territorial structures of authority (such as those relating to feudal and jurisdictional rights or privileges) and enclaves and spaces of ambiguous or shared sovereignty, which would be exchanged or redistributed between neighboring states. To a large extent, such practices were possible thanks to advances in cartography, geodesy, and topography. These not only made it possible to measure, describe, and represent territory with scientific precision, but also contributed to imagining the state as a unitary, continuous, and homogeneous geometric space (Biggs 1999; Branch 2014; Solans et al. 2019, pp. 30–35). Hence, the prominent role played within boundary commissions by expert bodies provided with the necessary technical capacity to adequately undertake these tasks.

Handbooks of international law published in the first half of the nineteenth century in countries under liberal governments provide a good indicator of the progress and consolidation of territorial conceptions of sovereignty and the nation (where the latter is understood as the political association of citizens within a territory and the sole subject of sovereignty), as well as of the role ascribed to border delimitation processes as essential tools in order to make them effective. With regard to the Iberian case, it is worth mentioning as examples the works of José María de Pando (1843) and Antonio Riquelme (1849) in Spain, as well as those of Vicente Paiva (1850) and Antonio Lobo (1853) in Portugal. Among these, it was De Pando (a Spanish–Peruvian liberal politician and diplomat) who devoted the most attention to these issues and who most clearly summarized the aforementioned conceptions. Firstly, De Pando defined the state or nation in the following terms:

> A certain number of men and families who, having gathered in a country and established their dwelling therein, come together and submit to a common leader, with the aim of jointly ensuring security for all, form a state or nation. The state is therefore a society of individuals, whose purpose is the preservation and happiness of its members; governed by positive laws emanating therefrom; and having ownership over a portion of territory (De Pando 1843, p. 59).

As for sovereignty, according to De Pando,

> In a broad sense, [it] consists of the set of rights of an independent state in relation to its purpose: it involves, 1. full independence of the state in relation to foreign nations; 2. the legitimate power of the government, or the authority that the purpose of the state requires (De Pando 1843, p. 59).

But, as the author points out, it is not merely that ‘the jurisdiction of a state cannot be exercised except within its own territory’ (De Pando 1843, p. 148); what is more, ‘the territory is the most inviolable of national assets: since in the absence of said inviolability, persons and the goods of individuals would be in danger at every step’ (De Pando 1843, p. 110). The definition of the national (or state) territory put forward by De Pando indeed calls upon exclusive control over it as a fundamental principle of sovereignty:

> A nation’s territory is the portion of the surface of the globe of which it is the owner and over which its sovereignty extends. What some have termed the state’s right to property (*jus in patrimonium reip.*) amounts to the power to exclude all foreign states and individuals from the use and appropriation of the territory, and of all things situated therein (De Pando 1843, p. 100).

And, following this line of reasoning, the precise delimitation of borders is deemed to be an essential task in order to maintain peace between neighboring countries and, therefore, the state’s independence and its ability to guarantee the safety of persons and goods:

> To establish as exactly as possible the limits or boundaries of the respective territories is an endeavour of the utmost importance for all nations, in order to prevent disputes and even wars that have often arisen from uncertainty (De Pando 1843, p. 107).

2.3 The Theoretical Foundations of the 1864 BT: An Overview of the General Instructions and Delimitation Criteria of the JBCs

The background and the process of preparation and execution of the 1864 BT have been examined in detail in some works (especially Barreiros

1961–1965; Godinho 2011; Cairo and Godinho 2013; García-Álvarez and Puente-Lozano 2015). By way of a summary, it is worth noting that in the mid-nineteenth century, certain areas of the *Raya* that lacked a precise delimitation continued to be the subject of frequent and sometimes violent disputes, of mainly two types: on the one hand, between the inhabitants of localities on both sides over the possession or use of certain resources (mainly water, pasture, and firewood); and, on the other, between these inhabitants and the agents of authority representing both kingdoms, who in some cases had to face not only the troubles arising from said disputes but also the problem of smuggling. In June 1854, the Portuguese government proposed to the Spanish government the creation of 'a Joint Commission in order to attempt to examine what is the true dividing line that must be established and definitively drawn up' (AHN, doc. 1).

Following long and complex negotiations, the JBC in charge of preparing a Boundary Treaty from the mouth of the Minho River to the confluence of the Caia and Guadiana rivers, attached to the Ministry of State (in the case of the Spanish section) and the Ministry of Foreign Affairs (in the case of the Portuguese section) was formally established in the town of Vigo on September 9, 1855. From then on, this commission, which was of a fundamentally technical nature, carried out the work of identifying and provisionally delimiting the border line, which went on until 1863. Once this phase was concluded, the governments of both countries created a new JBC, diplomatic in nature, which was tasked with negotiating and finalizing the definitive text of the 1864 BT, as well as its annexes and the general instructions to execute the demarcation. These were finally approved in November 1866. As already mentioned, demarcation work would in fact go on until 1906 and took place over two periods (1867–1873 and 1882–1906), separated by an interval of almost ten years during which the JBC were inactive.

Although they were not expressed through systematic arguments or by means of specific and explicit definitions, the conceptions about the border and how to delimit it shared by the Spanish and Portuguese governments and other central institutions which were responsible for the 1864 BT are reflected in the files and reports of the Spanish Ministry of State and the Portuguese Ministry of Foreign Affairs (drawn up by officials, referred to as *negociados de límites* in Spain, entrusted with these tasks), as well as in the minutes and reports of the JBCs. They also feature in parliamentary speeches prior to the ratification of the 1864 BT and, of course, in the text of the Treaty itself. The documentation that gave rise to the process, which is now for the most part kept in the AHDMNE (Lisbon) and in the AHN (Madrid), is very extensive, and while part of the material preserved in Lisbon was published by João-Baptista Barreiros (1961–1965), much of it, and almost all of that preserved in Madrid, remains unpublished.

Having paid greater attention to the work of the technical phases of the JBCs in previous works (especially García-Álvarez and Puente-Lozano 2015, 2017a, 2017b), we approach this vast archival corpus by focusing on the analysis of the collections stored in Madrid that relate to two key diplomatic phases—the initial phase and the final phase—in order to understand the aims, methods, and general criteria of the delimitation process. We look first at the preparation of the instructions submitted to the JBCs before the start of the delimitation process, and then at the Treaty's final negotiations, which were conducted in Lisbon by the plenipotentiaries of both governments.

The instructions given to the JBCs were negotiated by the two governments between October 1854 and August 1855, setting the objectives and the general working methodology to be followed by the commission. Negotiations were based on a document drawn up by the Spanish government in October 1854 (AHN, doc. 2), on which the Portuguese government would make some proposals (largely additions) in April of the following year, which would be negotiated a posteriori between both parties until arriving at the final approved version (AHN, doc. 3). The first of these documents supports the need for a 'clearly established demarcation between the kingdoms of Spain and Portugal' as part of a treaty, since the bilateral treaties signed in the Middle Ages, such as that of Alcañices (1297), relating to the border between the crown of Castile and the kingdom of Portugal, did not contain any complete or precise delimitations of these boundaries. For their part, the so-called *Tombos* (descriptions of demarcation lines carried out unilaterally by Portuguese authorities since medieval times) had, according to the Spanish government, no probative value. Nonetheless, the document also specified that there was already a boundary delimited de facto by the demarcation of border villages, in keeping with centuries-old occupation and land use practices, in light of which Tamar Herzog (2015) has referred to the Portuguese–Spanish border as a 'frontier of possession'.

> There is no convention held by the respective governments which describes the line that separates their territories; the boundary is currently determined only by the villages' private boundary pillars, which serve both to pinpoint their particular property and indicate the sovereign jurisdiction of the two Nations (AHN, doc. 2).

On this basis, the instructions point to a demarcation procedure that required the JBCs members first to familiarize themselves with the sections of the border to be demarcated, and second to proceed according to the two types of possible situations. In those sections of the border where there were no boundary issues, the JBCs members were to 'establish the signs that are to demarcate the true international boundary, in the clearest and most stable way possible'. Its trajectory was to be recorded in acts drawn up by public notaries of both kingdoms, 'taking care to clearly express therein the names of the points through which the boundary runs and the locations where boundary markers or other demarcation signs are placed'. Conversely,

> In such places where there are claims between towns or differences of opinion between commission members, you shall restrict yourselves to reporting with the utmost impartiality to this Ministry on any doubts that arise and the grounds for disagreement, in order for these to be resolved at a higher level (AHN, doc. 2).

All of the above guidelines, in addition to others relating to expenses and the writing up of the minutes of JBCs' sessions, would remain in the instructions that were finally approved, which would also introduce—at the suggestion of the Portuguese government—some general criteria on how to delimit and resolve certain disputes and boundary issues deemed problematic. As such, the Spanish government readily agreed to the Portuguese government's request for Paragraph 1 to stipulate that in such sections where the border between the two countries already coincided with rivers, commission members should consider their 'alveus' (that is, their natural bed or channel) in order to establish the international dividing line, except when said alveus (a concept that was expressed by the term 'thalweg' in the Portuguese version of the instructions) could be seen to have a 'tendency to vary' (AHN, doc. 3; Barreiros, 1961–1965, I: I–II, doc. 10).

The Spanish government also agreed, albeit with nuances, to introduce an additional paragraph on how to delimit border lands whose use had hitherto been shared between the populations of both kingdoms, since these were the subject of frequent disputes. In keeping with the liberal view of property, the governments of both states saw the mere survival of such spaces as problematic, just like they introduced confiscation laws aimed at dismantling a large portion of the commons within their respective countries (such as the Madoz Law of 1855 in Spain, or the Portuguese Confiscation Law of 1861, which from 1869 would be applied to municipal and parish wastelands, albeit with exceptions) (Rueda and Espinha da Silveira 1993; Del Moral et al. 2007).

In order to address such situations in the border delimitation process, the Portuguese government proposed that these lands be equally distributed in all cases (AHN, doc. 4). However, the Spanish, having accepted the need to divide them, argued 'that no general rule for distribution can be established a priori and invariably, because in [these lands] their possession and enjoyment are not always identical, nor are the origins and the conditions' (AHN, doc. 5). Therefore, the wording that was finally approved stipulated that:

> The distribution of lands whose use is shared between the respective border settlements shall take place according to and in proportion to the rights of each party, in keeping with the title deeds that they present and the enjoyment that they benefit from; recognizing between them, in the absence of written documents, those [lands] of acknowledged and undisputed immemorial belonging, and this shall apply to any of the parties concerned (AHN, doc. 3).

The Portuguese government also proposed another general rule that the Spanish refused to accept: that 'boundary markers are to be placed on a point of solid ground, and should the terrain be mountainous, the boundaries must be established atop the peaks of the mountains and not on the slopes on the Portuguese side' (AHN, doc. 4). During negotiations conducted on this point in Lisbon in late June 1855 between the Portuguese minister of foreign affairs (the Viscount of Atouguia) and the Spanish ambassador to Portugal (Patricio de la Escosura), the latter argued, firstly, that the second part of the clause in question, stipulating that, in mountainous areas, the border be drawn along the line of peaks, was 'absolutely inadmissible', because:

> Right now, the aim is not, let's say, establishing or creating the boundary, but fixing clearly and definitely its line, so that the Joint Commission's task is reduced to declaring the limits in places in which they are clear and to fix them in places in which there are doubts, lack of clarity, or conflicting claims (AHN, doc. 6)

In response to this argument, the Portuguese minister clarified that the wording initially proposed by his government was confusing, in that it suffered from 'an error in its drafting' (AHN, doc. 6), and that what he actually intended was for the principle of the line of peaks to apply reciprocally, such that all northern and eastern slopes of the mountains along the border between the two kingdoms would remain Spanish, and in exchange all southern and western slopes would be Portuguese. In view of this, and as stated in the report sent to his minister of state on the proceedings of the negotiations, the Spanish ambassador argued that this type of criterion could not be negotiated without knowing exactly what territory each of the two states would gain and lose.

> I replied [to the Portuguese minister] that discussing such serious business was beyond my remit, and such is indeed the case. Only with comprehensive knowledge, which I lack, since there is not a single map of Spain and Portugal in this Legation, may negotiations be acceptable from this point of view; because, while there is no doubt that disputes and even ongoing struggles between border villages would thus be rooted out, which is the apparent and sincere objective of the agreed demarcation, one must also examine to what extent it may or may not be beneficial for Spain to establish a principle, by virtue of which it could lose some portion of its current territory, without knowing what it would receive in return (AHN, doc. 6).

A second key phase for understanding the general theoretical principles that guided the delimitation agreed upon in the Lisbon Treaty is its final diplomatic negotiations. Although these had in fact begun in 1862, both delegations cut short their work to gather from two technical members of the JBC (the Portuguese military engineer Alexandre de Vasconcelos e Sá and the commander of the General Staff of the Spanish Army, José de Castro-López) new reports, maps, and more detailed plans on the territorial issues to be settled in the Treaty, which were not completed until mid-1863 (García-Álvarez and Puente-Lozano 2015, pp. 86–88). It was at this point that the diplomatic JBC that would conclusively negotiate the Treaty was created, with its meetings held between the end of 1863 and September 1864 in Lisbon. The Spanish section of this commission was made up of the Marquis de la Ribera, Spain's ambassador to Lisbon, acting as president, and the diplomat and politician Facundo Goñi, acting as secretary; while the Portuguese section was made up of the Marquis and Duke of Loulé, the minister of foreign affairs, acting as president, and Jacinto da Silva Mengo, an official from said ministry, acting as secretary.

The diplomatic JBC that negotiated the 1864 BT thus built on the extensive work carried out by the technical JBCs from 1855 onwards, essentially prepared by the military engineers and officers of the General Staff of the Army attached to those commissions. Based on this documentation, the diplomats and ministerial officials who advised them and gave them instructions from Lisbon and Madrid were able to learn in some detail about the population, the area, the land uses, and the agricultural quality of the disputed lands, which were key parameters for estimating the value of said lands, as well as the needs of the local communities involved in these disputes. In their arguments and deliberations, the various parties made use of sketches, plans, and maps, and quite frequently used the aforementioned statistical data and the reports produced during the direct survey of the border regarding the importance of the disputed resources for the populations concerned. In this regard, it is worth highlighting the role of inter-scalar bridging or mediation played by the technical JBCs deployed on the ground in attempting to reconcile, as far as possible, the interests of the central powers and those of local communities, or, in other words, the general theoretical principles and the particular empirical realities—a vital task in order to ensure the successful implementation of nineteenth-century boundary treaties, both in the case of the *Raya* and in other Iberian and European cases (Di Fiore 2017, García-Álvarez and Puente-Lozano 2017b).

In addition, the reports of the officials of the Ministry of State and of the two aforementioned Spanish plenipotentiaries who, as members of the diplomatic JBC, ultimately negotiated the 1864 BT frequently refer to three criteria that should ideally guide the delimitation agreements in those disputed territories (see, e.g., AHN, doc. 7, Instructions of the Spanish government of 30 November 1862 and 20 August 1863; and Reports by the plenipotentiaries of 27 February 1864). These three criteria were: *regularity and clarity in the geographical outline of the border* (ensuring that, wherever possible, it followed natural and permanent boundaries, as well as straight lines instead of curved or winding lines); *equity* (in the allocation or exchange of disputed territories between both countries); and *justice* (understood as respect for existing title deeds or proof of possession, as well as the commitment to satisfy the 'legitimate' and 'vital interest' needs of border local communities, thus ensuring that the line of partition that was to be approved would not threaten their economic survival). To these criteria, the Portuguese plenipotentiaries added the need to adopt solutions that their government could present as fair and equitable before parliament, the press, and public opinion (AHN, doc. 7, reports of 27 February and 10 May 1864), an argument that the Spanish would later take up too (AHN, doc. 7, report of 14 April 1864).

The latter argument came to the fore above all (albeit not exclusively) during the resolution of issues that were particularly sensitive or controversial from a political point of view, such as those involving the Couto Mixto and the 'promiscuous villages', which had a bearing on sovereignty over entire populations, and which we shall refer to in the following section. Moreover, it was frequently accompanied by the idea that, in order to guarantee the principle of equity, the 'cession' of a disputed territory to the other country must as far as possible involve compensation, an exchange, or be offset by the beneficiary country through the transfer of another disputed territory of similar value to the ceding country. For the purposes of this type of operation, and in the absence of official and modern censuses and topographic maps of many of these border territories, the surveys and plans conducted on the ground by the technical JBCs' experts acquired decisive importance in the negotiation process.

2.4 Territorial Issues and Conflicts Along the Portuguese–Spanish Border: Typology and Solutions Approved in the 1864 BT

The 1864 BT devotes its first twenty-three articles to qualitatively describing the border to which it applies, from the mouth of the Minho River to the confluence of the Caia and Guadiana rivers (United Nations 1991, p. 275–283). As this line is described, the Treaty specifies the agreed solution for the territorial issues or disputes addressed therein, located for the most part in the Galician–Portuguese sector of the *Raya*, and particularly in the so-called *Raya Seca* (i.e., Dry Borderline) which straddles the Spanish province of Ourense and the Portuguese districts of Vila Real and Bragança: no fewer than seventeen of the twenty-five issues settled in the 1864 BT are located along this section, of approximately 225 km in length (Fig. 2.1).

The nature of the issues addressed in the 1864 BT was not uniform, nor were the solutions adopted in relation to them. As such, it is possible to differentiate between two main types of issues: on the one hand, those concerning the possession and use of agro-silvo-pastoral lands and resources by neighboring border communities, which were disputes of a purely or primarily local nature, and which, depending on what form they took, were referred to in the BT as 'disputed lands' (in Spanish: *terrenos cuestionados*) or 'dubious lands' (*terrenos dudosos*); and, on the other hand, those relating to the Couto Mixto and the promiscuous villages, which, in the minds of the diplomats who negotiated the BT, were the only ones to have a truly international significance, since they affected the ability of the states to fully exercise their sovereignty (AHN, doc. 7, 27 February 1864). The preamble to the 1864 BT indirectly references both types of problems, supporting the need to establish this diplomatic instrument:

> Having taken into consideration the state of uneasiness existing in many towns situated on the confines of the two kingdoms owing to the inexistence of any precise delimitation of territory or any international treaty defining such a delimitation; desiring to put an end once and for all to the unpleasant conflicts which, on that account, arise at various points on the frontier and to establish and consolidate peace and harmony between the border populations, and, finally, recognizing the need to end the anomalous situation in which, under the shadow of ancient feudal traditions, some towns bordering on the dividing line of the two States have remained until now, to the manifest and common disadvantage of the said States [...] (United Nations 1991, p. 275).

We shall now briefly examine both issues, as well as the solution provided to each in the Treaty.

2.4.1 Disputed and Dubious Lands

Most of the boundary issues addressed by the 1864 BT had to do with agro-silvo-pastoral lands that were either disputed by border communities ('disputed lands', as they are called in the Treaty), or were traditionally used jointly, under common agreement ('dubious lands' or 'lands of doubtful ownership'). No matter how small these may have been in most cases, the control and use of these spaces, which included some islands along the course of the Minho River, were vital in the context of the subsistence economies of local border communities. In some sections of the border, such as the aforementioned *Raya Seca*, the case files put together in preparation for the BT, or with regard to so-called 'territorial violations' associated with the Portuguese–Spanish border during this period of the nineteenth century, attest to a heightened potential for conflict between border communities, as well as between these communities and the fiscal agents of both kingdoms, linked to the uncertain nature of boundaries in the lands at issue. This potential for conflict was due in large part to the growing demographic pressure on resources (particularly on the Spanish side), but also to the growing activity of fiscal agents. This resulted, for instance, in numerous seizures of livestock or the arrest of residents when they were believed to have committed a 'territorial violation' by illegally crossing the border without paying compulsory customs duties (AHN, doc. 7, report by Fidencio Bourman, president of the Spanish section of the JBC, dated 6 April 1862).

The very activity of the JBCs during the border survey process prior to the 1864 BT, which involved the delimitation of some lands under discussion, intensified altercations with and between local communities, including the destruction or shifting of boundary markers provisionally established by the JBCs, as occurred in the case of the disagreements between the Spanish villages of Terroso, Soutochao, and Berrande and the Portuguese villages of São Vicente and Segirei. Some of these issues, such as those relating to Monte da Madalena (between the Spanish villages of Compostela, Lodeiros, and Manín and

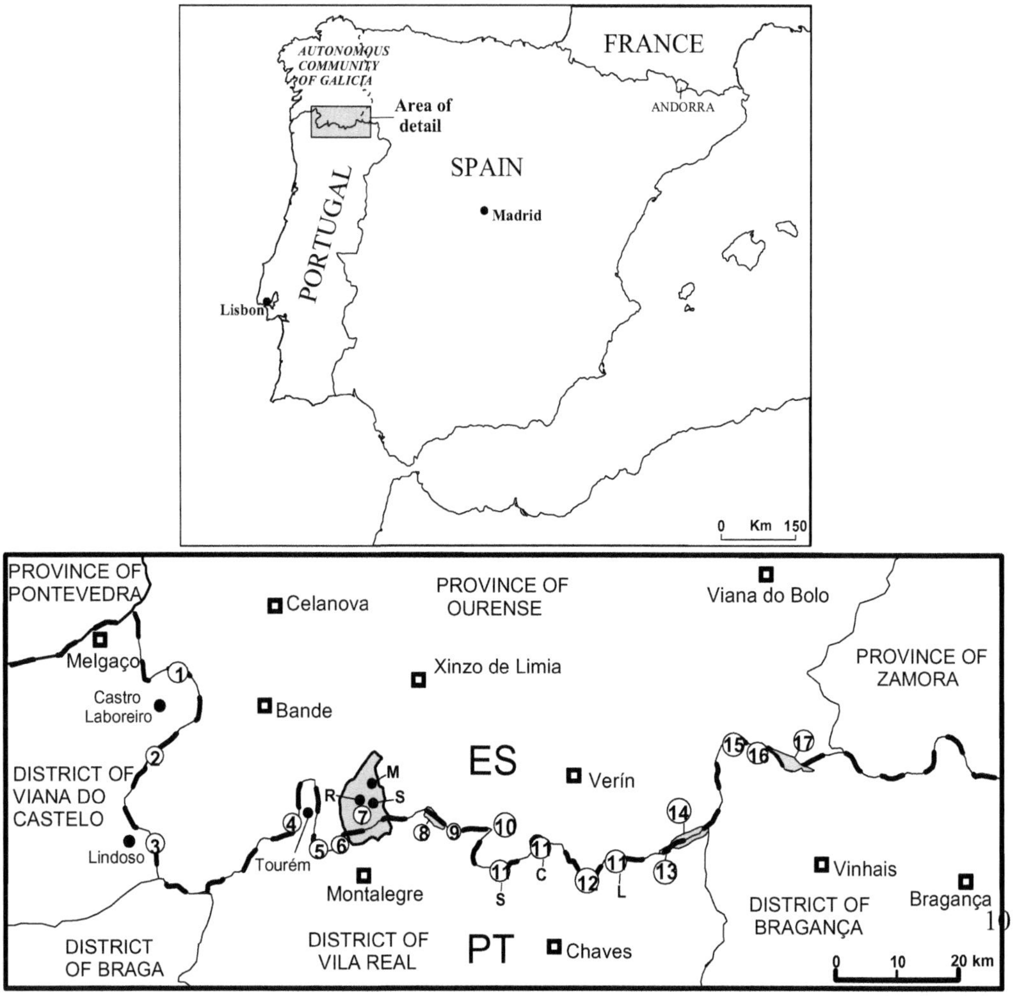

Fig. 2.1 The Spanish–Portuguese *Raya seca* (Dry borderline) between the Ourense province and the Vila Real and Bragança districts. Territorial issues addressed in the 1864 BT negotiations. The numbers on the bottom map represent the following issues (ES: Spanish villages; PT: Portuguese villages): 1. Gorgua (ES) vs. Adofreire (PT); 2. Pereira (ES) vs. Amexoeira (PT); 3. Monte da Madalena; 4. Guntemil and Requiás (ES) vs. Pitões das Júnias (PT); 5. Vilariño (ES) vs. Tourém (PT); 6. Vilar (ES) vs. Sabuzedo (PT); 7. Couto Mixto (R: Rubiás; S: Santiago; M: Meaus); 8. A Xironda (ES) vs Santo André (PT); 9. A Xironda (ES) vs Vilar de Perdizes (PT); 10 Road from Videferre to San Millán; 11. "Promiscuous villages" (S: Soutelinho; C: Cambedo; L: Lamadarcos); 12. Land portions where the rivers Támega and Pequeno did not formed the boundary; 13. Terroso (ES) vs. São Vicente (PT); 14. Soutochao, Berrande and Tomonte (ES) vs. Aveleda, Segirei and Vilaseco (PT); 15. Barxa (ES) vs. Cisterna (PT); Esculqueira (ES) vs. Pinheiro Novo (PT); 17. Chaguazoso and Manzalvos (ES) vs. Pinheiro, Vilarinho, Cerdedo, Casares and Carballas (PT). The dashed black line represents the bounday approved in the 1864 Treaty. Towns that were head of judicial districts (*partidos judiciales*) in Spain and of municipalities (*concelhos*) in Portugal are represented by rectangles. *Source* Own elaboration on the basis of José de Castro-López's sketch map of the border (*Croquis de una parte de la Frontera entre la Provincia Española de Orense y Portugal*, July 1863, ACEGCGE) and AHN (doc.8)

the Portuguese village of Lindoso), which concerned an area of about 1300 ha; or that involving the Spanish village of Chaguazoso and the Portuguese villages of Pinheiro Velho, Vilarinho, and Cerdedo, over a 600 ha stretch of land, gave rise to frequent altercations, some of which were bloody.

But apart from dividing up these 'disputed lands', the 1864 BT—in accordance with the provisions of the additional paragraph of the instructions approved in 1855, and in keeping with the liberal conceptions already mentioned—would also provide for the elimination and distribution of almost all lands that were undivided or of shared use between villages on either side of the border.

For the JBCs, and as stated in the BT itself, such lands were in fact considered 'dubious lands' or 'lands of doubtful ownership' (in Spanish: *terrenos de pertenencia dudosa*), even when their joint use was peacefully conducted under agreements between the various towns. This was the case, for example, of the so-called *Reyertas* (three pieces of land that together totalled about 1750 ha, and that for centuries had been enjoyed by the Spanish municipality of Alburquerque, in the province of Badajoz, and the Portuguese towns of Ouguela and Arronches, in the Portuguese district of Portalegre).

> Despite the fact that no quarrel exists in the shared enjoyment of the pastures—the Spanish plenipotentiaries thus pointed out with reference to a land of doubtful ownership between the Spanish village of Esculqueira and the Portuguese villages of Pinheiro Novo and Pinheiro Velho—, we have deemed it preferable to divide it up rather than to leave it in an uncertain state that may in the future provide room for disputes between the two villages, something that is always unfavourable to the interests of both countries (AHN, doc. 7, 27 February 1864).
>
> By agreeing to cede the aforementioned portion as property—argued the same envoys to justify the cession to Portugal of the so-called 'joint pastures' (*pastos mixos*) belonging to the Couto Mixto—, our decision has been influenced by the firmly held belief acquired in studying these matters that cases of usufruct between border towns are usually very harmful. Usufruct almost invariably tends to give rise to clashes and altercations: under its cover, smuggling is carried out with impunity despite the vigilance of guards, and, in short, usufruct rights render illusory many of the effects of international delimitation (AHN, doc. 7, 27 February 1864).

The 1864 BT would therefore undertake the partition of disputed and doubtful lands, either in equal parts, as in the case of those covered by Articles 2, 3, 4 (in connection with the issue of Monte da Madalena), 5, 8, 15, 19, and 22 (relating to the aforementioned *Reyertas*), or by means of other criteria, such as those applied to the islands on the Minho River (Article 1) or to the lands described in Articles 13 and 14. In the case of the Minho River, and as we have studied in detail elsewhere (García-Álvarez and Puente-Lozano 2017a), the Treaty established that, given the river's shifting course (mainly due to the numerous fisheries built on both of its banks), the border should be drawn so that it 'shall proceed along the centre of the principal stream of the Minho to the confluence of the river Barjas or Troncoso'. This expression was clarified in Annex I of the Treaty, which contains the regulations applicable to boundary rivers (United Nations 1991, p. 284–287), specifying that the border be drawn along the median or equidistant line between both banks of the river. Furthermore, albeit in keeping with this principle, it was agreed that the disputed river islands would be allocated, respectively, according to their greater geographical proximity to the Spanish or Portuguese banks. This criterion was in fact motivated by strategic and military reasons (to facilitate the defense of each kingdom and to more effectively control smuggling). As for the case of the lands described in Articles 13 (which covers the aforementioned issue of Chaguazoso) and 14, these were divided up between both kingdoms in proportion to the sizes, needs, or title deeds of the villages concerned.

The 1864 BT also settled some boundary issues through exchanges, for instance granting to Portugal the land disputed by the Portuguese village of Ameixoeira and the Spanish village of Pereira (Article 3) in exchange for Spanish sovereignty over the land disputed by the Portuguese village of Tourém and the Spanish village of Vilariño (Article 6). Similarly, also through exchanges, it eliminated some small enclaves or other types of territorial anomalies that the JBC saw as 'irregularities' that could trigger conflicts. In this regard, for example, the Treaty put an end to the problem arising from the road that connected the Spanish villages of San Millán and Videferre, a small section of which (known as 'a Ponte Nova') crossed through Portuguese lands, which led to frequent incidents between the residents of these villages and the Portuguese fiscal guard. Under Article 9, the territory through which this road passed was to belong entirely to Spain, though in exchange for this cession, located in the province of Ourense, Article 20 would grant Portugal sovereignty over the dubious land that covered the Tapada del Manso and the Dehesa de la Genestrosa, on the border of the Spanish province of Salamanca. Two further provisions slightly modified the border on the banks of the Támega and Pequeno rivers (Article 11) and the Caia River (Article 23), where, perhaps due to natural shifts to their course, there were some portions of land that interrupted the border formed by these rivers. Under the aforementioned articles, and without modifying the ownership of the private landlords of these portions, sovereignty over them was exchanged for that of other lands, thus ensuring that the border fully coincided in said sections with the course of the river.

2.4.2 The Couto Mixto and the Promiscuous Villages

Over the last twenty years, the history of the Couto Mixto (as it was called in Galician), and, to a lesser extent, that of the promiscuous villages, has given rise to an extensive literature, some of which has examined in detail the process that led to their elimination, as approved by the 1864 BT (García-Mañá 2005; Godinho 2011; García-Álvarez 2019a). Also situated along the Galician–Portuguese *Raya Seca*, the two territories had different characteristics, although some of the problems that they raised were shared, which is why they were negotiated together under the Treaty.

The Couto Mixto (Portuguese: *Couto Misto*; Spanish: *Coto Mixto*) was, prior to its elimination, a territory with an area of about 25 km^2 that lay to the south of the province of Ourense, in the Salas valley. In the mid-nineteenth century, it was home to some 250 households (around 1000 inhabitants), distributed across three villages (Santiago, Rubiás and Meaus), who essentially relied on livestock for their livelihoods (Fig. 2.2). By virtue of privileges initially acquired under a 1518 judgment, issued as a result of a jurisdictional dispute between the Houses of Braganza (in Portugal) and Monterrey (in Spain), this territory in many respects retained an independent status from Spain and Portugal. Consequently, the inhabitants of the Couto Mixto could, for example, freely adopt the nationality of their choosing (Spanish or Portuguese) and easily change it at their convenience; they barely paid taxes; men there were exempt from military service; they had their own judge chosen by the residents; they could grow or trade any product without restrictions; they benefited from the right of asylum for those fleeing justice (except for those persecuted for blood crimes); and they could freely transport any product to the neighboring village of Tourém, in Portugal, via a 'privileged road' about 6 km long, where neither country's fiscal agents could act. Historian Isabel Braga (1997) has referred to this territory as a remnant of the 'diffuse frontier' characteristic of the Middle Ages and the Ancien Régime, riddled with areas of confused sovereignty. For her part, Herzog (2015, p. 230), using a suggestive metaphor, has described it as a set of 'islands of jurisdiction in a sea of mixed land'—a reference to the fact that, at least until its extinction, the Couto Mixto's dependence vis-à-vis Spain or Portugal was exclusively associated with the residents and their homes, not with the territory of the villages themselves, whose assignment was unknown or was left out of this system of dependencies.

As for the promiscuous villages (Spanish: *pueblos promiscuos*; Galician: *pobos promiscuos;* Portuguese: *povos promíscuos*), these were made up of three villages (Soutelinho, Cambedo, and Lamadarcos) that were geographically

Fig. 2.2 Detail map of the *Couto Mixto*, included in the *Croquis de una parte de la Frontera entre la Provincia Española de Orense y Portugal*, at a scale of 1:100.000, drawn by the Spanish Commissioner José de Castro-López (March 1863). We have highlighted with a red dashed line the so-called privileged road (in Spanish: *camino neutral* o *privilegiado*), which linked the territory of the Couto (highlighted in cream color) with the Portuguese village of Tourém (in Spanish: Torey), and with a black dashed line the boundary line agreed in the 1864 BT, which abolished that road and granted Portugal sovereignty over a small southern sector of the Couto. *Source* ACEGCGE, Ar. I-T. 6-C.1–15 and 16

intersected by the border, which divided their limits, their centers, and even some of their streets and houses (Fig. 2.3). The three villages together came to a total of little more than 19 km^2 and a population of close to 231 households (around 1000 inhabitants), 162 of whom lived in the Portuguese sections and 69 in the Spanish sections. Unlike the Couto Mixto, in the case of the promiscuous villages the territory that belonged to each kingdom was clear. The persistence of the Couto Mixto and the three promiscuous villages openly fostered smuggling

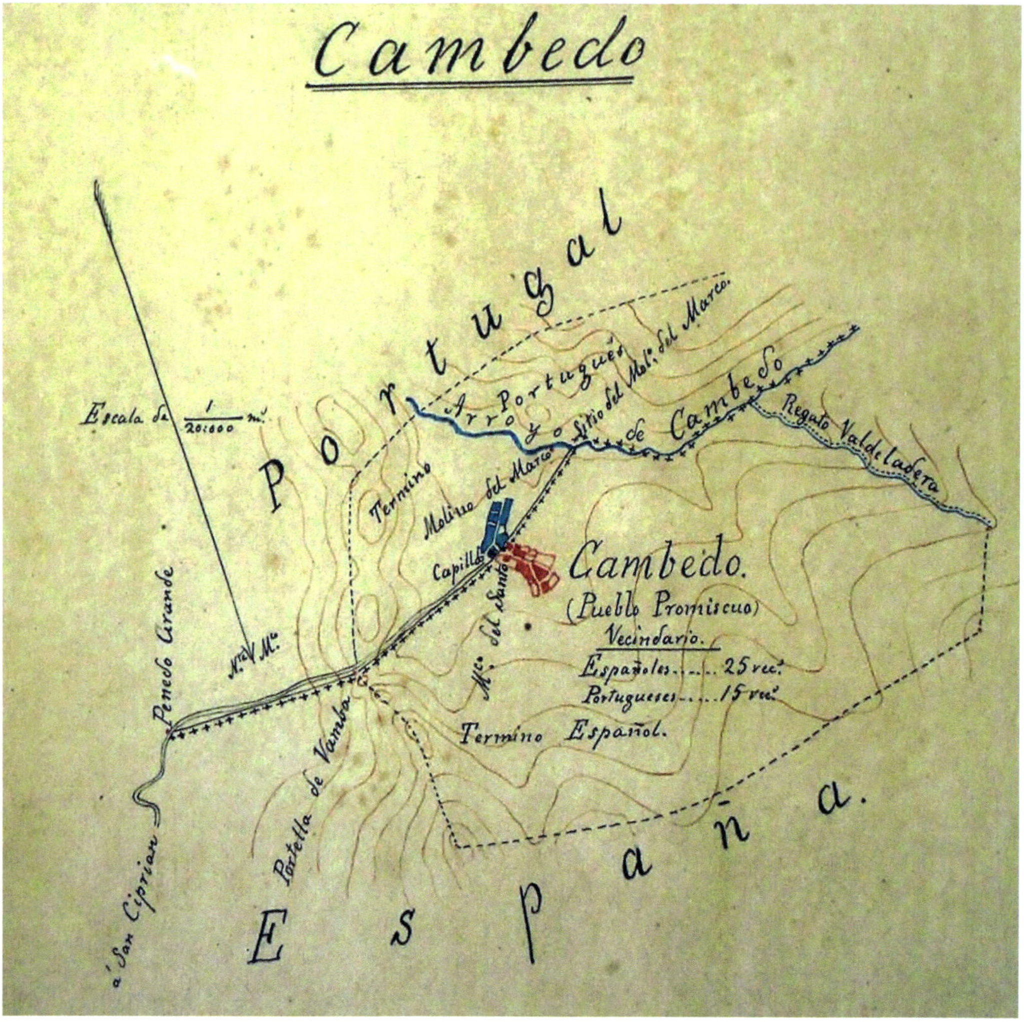

Fig. 2.3 Sketch map of the 'promiscuous village' of Cambedo, at a scale of 1: 20,000, by José de Castro-López (July 1863). As can be seen in the figure, before the approval of the 1864 BT, which gave Portugal sovereignty over the entire promiscuous villages, the border line divided their respective territories and population centers. As the cartographer pointed out on the map, in the case of Cambedo 25 households (around 100 inhabitants) were located on the Spanish side of the village, here colored in red, whereas 15 households (around 60 inhabitants) were on the Portuguese side, here colored in blue. *Source* AHDMNE, Cartas Comissão de Límites, Sig. 14-C

while undermining the effective enforcement of the principle of sovereignty.

For its part, the Couto Mixto was viewed from the liberal perspective as an intolerable feudal relic. As a matter of fact, it is precisely the issue of the Couto Mixto that gave rise to more abundant arguments and that gives us a better picture of the concept of sovereignty shared by the Spanish and Portuguese liberal governments, which jointly invoked the principles of exclusivity, administrative efficiency, national security and territorial integrity, and the rule of law to justify the need to put an end to the special status of this territory. Thus, for example, a recurring argument used by the central powers was that of upholding the right of both countries to control their territory and establish order and the rule of law by means of effective borders. From this perspective, and as stated in the memorandum sent by the Spanish government to the Portuguese government in November 1858 in an attempt to settle the issue:

> The existence of the Coto Mixto under its current circumstances is impossible [because] it produces evident damage to sovereign rights and significant harm to fiscal interests; [which is why the government] wishes to eliminate the illegal status of the villages in the Coto Mixto, replacing the anarchy that exists therein with the rule of Spanish law (AHN, doc. 8).

By the same token, from the Spanish perspective (albeit not from the Portuguese perspective), upholding the principle of territorial integrity was of great importance, since it was considered essential not only from the point of view of sovereignty, but also from that of national security. Not surprisingly, in the Spanish take on the matter, the Couto Mixto was regarded and presented as an enclave surrounded by Spanish territory (although this was not in fact the case on its southern side), linked to Portugal by a road along which border guards were unable to act: a geographical argument that, according to the Spanish stance, had to prevail over any other consideration, as summed up by Antonio Benavides, minister of state at the time, who upon presenting the 1864 BT to the Spanish Senate explained that the Couto Mixto was 'unquestionably within Spain for the very simple reason of having encroached [...] by more than a league into Spanish territory' (DSCS, 6 February 1865: 318).

But, what is more, the Couto Mixto was perceived as setting a 'bad example' for the surrounding towns, and more precisely as an anachronism that, given its special privileges, encouraged smuggling, provided refuge to enemy factions and armies, as well as those fleeing justice, and in practice dissolved or destroyed the effectiveness of the border with the neighboring country. As observed by the Spanish plenipotentiaries in charge of the final negotiation of the BT:

> At issue here is a group of three villages, namely Santiago, Rubiás, and Meaus, a kind of mini-republic, which since time immemorial have led an independent, irregular, and anomalous existence, already incompatible with the conditions of the present time; which cover a territory of two thousand five hundred and five hectares, surrounded by Spain; which, given their exceptional status, set a harmful example to the surrounding Spanish villages; and, finally, which destroy, politically and economically, the border between both kingdoms (AHN, doc. 7, 27 February 1864).

After arduous and lengthy negotiations, which, unlike what occurred in many of the disputed lands, did not take into account the opinions of the inhabitants, the 1864 BT resolved, on the one hand, to transfer practically the entire Couto Mixto (including its three constituent villages) to Spain, with the exception of the 'joint pastures' (*pastos mixtos*), a stretch of land of around 330 ha located in the Couto Mixto's southern section, which, in response to requests by some neighboring villages, came under Portuguese sovereignty (Article 7); and, on the other hand, in exchange for the decision regarding the Couto Mixto, to transfer to Portugal all three promiscuous villages, which, in the case of Soutelinho, involved extending the border with Spain to an area 90 to 100 m in width contiguous to the settlement (Articles 10 and 11). Furthermore, Article 27 established the possibility for inhabitants of the Couto Mixto villages who were 'really Portuguese subjects', as well as those of the promiscuous villages who were 'really Spanish subjects', to 'have the right to retain their nationality if it so suits them', for which purpose, 'both [...] shall declare their decision to the local authorities within a period of one year, reckoned from the date on which the present Treaty is put into operation' (United Nations 1991, p. 282). The formal handover of the Couto Mixto to the Spanish authorities took place on June 23, 1868, and that of the promiscuous villages to the Portuguese authorities on 27 June, without any notable incidents.

2.5 Concluding Remarks: Modern State Sovereignty as Geometric Purification of Territory

The 1864 BT provides an exemplary reflection of the conceptions of sovereignty characteristic of liberal states and, more broadly, of the consolidation of the territorial state in the Iberian

Peninsula. In contrast with the idea of sovereignty still in force in absolute monarchies, of a dynastic, hereditary, and jurisdictional nature, based on relations of loyalty and vassalage between the sovereign and his subjects, the liberal governments behind the Treaty based the sovereignty of the nation-state on the exclusive and indivisible jurisdiction and control over its territory, delimited precisely by linear borders.

The 1864 BT did not create the linear border per se, because, as the initial instructions to create the JBC observed from the outset, it had already been established de facto by the boundaries between local communities, most of which would be adopted by the Treaty. But it did provide this border with an international legal status and, above all, completed and specified it, enabling the continuous and discrete delimitation of the state territory through a series of operations that, together, amounted to a kind of geometric and geopolitical purification. This purification removed the impurities or 'irregularities' (the term usually used in the process) that, in the view of both governments, gave rise to troubles among border communities and hindered the effective enforcement of state sovereignty, such as lands of doubtful, shared, or disputed ownership; villages and islands intersected by the border line; territories of hybrid or diffuse sovereignty, or even semi-independent of both states; enclaved roads and territories; and, finally, the river boundaries along which the wet border had been partially interrupted by changes to their course.

As such, wherever there were disputed lands, or undivided lands of shared use or doubtful ownership, as was the case with many territories along the *Raya*, the Treaty undertook to partition them. Wherever, as in the promiscuous villages, the border, albeit precise, ended up splitting the limits or center of a settlement, thereby encouraging smuggling and eroding or hindering the effective enforcement of the principle of sovereignty, this line was modified and shifted so as to ensure that these limits fell fully under a single country, in this case Portugal. As for the case of the Couto Mixto, where sovereignty was shared between the two states or, in certain respects, was exercised independently of the states (true to a typically pre-modern a territorial and jurisdictional model), the agreed solution would be the annexation of its territory to the two countries, transferred for the most part to Spain. Finally, in order to guarantee the complete continuity of the territory that belonged to each state, the Treaty also abolished any roads and lands enclaved in either kingdom, exchanging them via swaps. Invoking the argument that any river borders should follow the course of the waters, in order for these to be clear and natural boundaries, it rectified certain sections to skirt any remaining portions of land that may act as a dividing line between the two kingdoms.

The very representation of the Couto Mixto as an enclave, whose possible transfer to Portugal could threaten its territorial integrity, promoted by the Spanish government in particular, despite it not being true in geographical terms, perfectly fits the logic of geometric and geopolitical purification of space that underpins the territorial state and the concept of sovereignty associated with it. As we have analyzed elsewhere (García-Álvarez 2019a), the extinction of the Couto Mixto sums up and illustrates to a greater extent than any of the other issues addressed by the 1864 BT the replacement of a jurisdictional border by a territorial one; that of diffuse or weak borders by linear, sharp, and hard ones; and that of hybrid and shared sovereignties by exclusive ones. Such developments had in many cases been brewing since at least the time of the enlightened monarchs, but they did not materialize or take full form until the establishment of the liberal state.

As has been shown in this and other works (e.g., García-Álvarez and Puente-Lozano 2015; García-Álvarez and Puyo 2015; García-Álvarez 2019b), the preparation and subsequent execution of the 1864 BT also revealed the importance of geographical, statistical, and cartographic knowledge in a dual sense: on the one hand, as a source of support for arguments and claims by either of the parties, and to assist with diplomatic decisions regarding the border outline; and, on the other hand, as an expression and a vehicle of a geometric kind of territoriality, necessary for imagining each state as a unitary, continuous, homogeneous, two-dimensional, Euclidean, and clearly delimited space. Also in this regard, one could argue that the 1864 BT reflected a general trend in European diplomacy since the start of the nineteenth century in the manner of describing and representing the subjects and the spatial limits of sovereignty: in contrast with the territorial descriptions characteristic of Ancien Régime peace or 'border' treaties (made up, until well into the eighteenth century, of lists of settlements, fortresses, lands and jurisdictions that fell under the sovereignty of each monarch), the nineteenth-century boundary treaties expressed sovereignty in terms of delimited and cartographically represented territories.

Indeed, in the case at hand, this way of representing sovereignty would be made explicit and developed especially during the demarcation phase of the Treaty of 1864, regulated by a set of instructions approved in November 1866 (*Instrucciones para ejecutar la demarcación de la frontera entre España y Portugal*) and culminated with the General Act of Demarcation, approved, as mentioned above, in December 1906 (*Acta general de demarcación de la frontera entre España y Portugal desde la desembocadura del río Miño hasta la confluencia del Caya con el Guadiana*. Reproduced in *Acuerdos fronterizos*... 2005, pp. 217–333). This lengthy demarcation process would involve the production by the JBCs of a scientific cartography of the boundary line approved under the Treaty, drawn up at various scales, as well as detailed 'geometric descriptions' thereof, which form the main text of the partial demarcation acts and of the General Act that compiles them, and that provides an account of the trajectory followed by this line along the 801 boundary markers that materialize its demarcation (García-Álvarez and Puente-Lozano 2015, pp. 87–91). In these descriptions, references to significant features in the landscape (rivers and streams, relief features, roads, houses, settlements, etc.) are accompanied by quantitative geometric and topographic parameters. These include the number of boundary markers, the distance in meters between them, and the astronomical bearing or direction (azimuth) of the line between each pair of consecutive boundary markers, measured in degrees.

As a final point, beyond its political dimension, the establishment of modern borders of course had important ramifications on many other levels, including the economic and the symbolic, and in the latter sense it was a crucial mechanism for the construction of national identities. But this is a matter that lies beyond the scope of this contribution, and we shall seek to address it in further research.

Archival Sources

ACEGCGE. Archivo Cartográfico y de Estudios Geográficos del Centro Geográfico del Ejército. Madrid.

AHDMNE. Arquivo Histórico-Diplomático do Ministério dos Negócios Estrangeiros. Lisbon.

AHN (Archivo Histórico Nacional/National Historical Archive, Madrid), doc. 1: Letter from the Portuguese ambassador in Madrid (Conde de Azinhaga) to the Minister of State of the Spanish government (Ángel Calderón) informing him of the interest of the Portuguese government in creating a Joint Commission to determine the border line between the two countries (AHN, Ext. Tr. 140, Neg. 207, Exp. 24).

AHN, doc. 2: "Instrucciones a los Comisionados de SM encargados de la demarcación de los límites en la frontera con Portugal" ("Instructions to the HM Commissioners in charge of the demarcation of the boundaries at the border with Portugal"), Madrid, 18 November 1854, signed by the Spanish Minister of State Francisco Pacheco (AHN, Ext., Tr. 138, Neg. 207, Exp. 20, doc. nº 123).

AHN, doc. 3: "Instrucciones para la Comisión Mixta encargada de la demarcación de la frontera entre los Reynos de España y Portugal, reformadas con arreglo a la correspondencia seguida al efecto entre los dos gobiernos" ("Instructions for the Joint Commission in charge of the demarcation of the border between the Kingdoms of Spain and Portugal, reformed in accordance with the correspondence followed to that effect between the two governments"), Madrid, 15 October 1855 (AHN, Ext., Tr. 138, Neg. 207, Exp. 20, doc. no. 72).

AHN, doc 4: Letter from the Spanish ambassador in Lisbon (Juan Bautista de Sandoval) to the Spanish Minister of State giving an account of the note of 25 April sent by the Portuguese Minister of Foreign Affairs (Viscount of Atouguia), Lisbon, 21 May 1855.

AHN, doc. 5: Letter from Spanish ambassador in Lisbon (Patricio de la Escosura) to Spanish Minister of State,

29 June 1855, reporting on his conference with the Portuguese Minister of Foreign Affairs and including some important indications made latter by G. Muro (AHN, Ext., Tr. 138, Neg. 207, Exp. 20, doc. nº 124).

AHN, doc. 6: Letter from Spanish Ambassador in Lisbon (Patricio de la Escosura) to the Spanish Minister of State, 28 June 1855, reporting on his conference with the Portuguese Minister of Foreign Affairs (AHN, Ext., Tr. 138, Neg. 207, Exp. 20, doc. nº 123).

AHN, doc. 7: Several documents regarding technical discussions and political instructions given by Spanish government to JBC Spanish commissioners Fidencio Bourman, Marqués de la Ribera and Facundo Goñi throughout 1862 to 1864 for resolving border territorial disputes in the framework of final 1864 BT diplomatic negotiations (AHN, Ext., Tr. 135, Neg. 207, Exp. 12).

AHN, doc. 8: *Memorándum adjunto al Proyecto de convenio sobre las Cuestiones de los Cotos Mixtos y pueblos promiscuos* (Memorandum attached to the Draft Convention on the Questions of Couto Mixto s and Promiscuous Villages), drawn up by the Spanish Government, November 1858 (AHN, Ext., Neg. 207, TR. 135, file 10).

References

Acuerdos fronterizos con Portugal y Francia. Colección de recopilaciones (2005). Ministerio de Asuntos Exteriores y de Cooperación, Madrid, vol. 1

Barreiros JB (1961–1965) Delimitaçao da fronteira luso-espanhola. O Distrito de Braga, I (I–II), 1961: 59–148: I (III–IV), 1961: 335–413; II (I–II), 1963: 83–170; II (III–IV), 1964: 81–176; III (I–II), 1964: 1–96; III (III–IV), 1965: 401–489

Biggs M (1999) Putting the state on the map: cartography, territory and European state formation. Comp Stud Soc Hist 41(2):374–405

Braga I (1997) A fronteira difusa entre Tras-Os-Montes e a Galiza ou as Povoaçoes místicas de Santiago, Rubiâes e Meâos. Brigantia 3–4:3–13

Branch J (2014) The cartographic state. Cambridge University Press, Cambridge

Cairo H, Godinho P (2013) El Tratado de Lisboa de 1864: la demarcación de la frontera y las identificaciones nacionales. Historia Política 30:23–54. Retrieved from: https://eprints.ucm.es/39822/

De Pando JM (1843) Elementos del derecho internacional. Imprenta de Alegría y Charlain, Madrid

Del Moral J et al (2007) Estado y territorio en España, 1820–1930. Los Libros de la Catarata, Madrid

Di Fiore L (2017) The production of borders in nineteenth-century Europe: between institutional boundaries and transnational practices of space. Eur Rev Hist 24:36–57. https://doi.org/10.1080/13507486.2016.1183594

DSCS: Diario de las Sesiones de Cortes: Senado. Cortes Españolas, Madrid

Flint C, Taylor P (2018) Political geography, 7th edn. Routledge, London

García-Álvarez J (2019a) Geopolíticas de la territorialidad moderna. El Tratado hispano-portugués de límites de 1864 y la extinción del Couto Mixto (1840–1868). Boletín Asociación Geógrafos Españoles 82(2798):1–36. https://doi.org/10.21138/bage.2798

García-Álvarez J (2019b) Territorialidad estatal y conocimiento geográfico. Reflexiones sobre la delimitación moderna de las fronteras ibéricas. In: Farinós J et al (eds) España: Geografías para un Estado postmoderno. Asociación de Geógrafos Españoles/-Geocrítica, Barcelona, pp 243–254. Retrieved from: http://www.ub.edu/geocrit/estadoposmoderno.pdf

García-Álvarez J, Puente-Lozano P (2015) Las Comisiones Mixtas de Límites y las representaciones geográficas de la frontera hispano-portuguesa (1855–1906). Revista Historiografía 23:67–100. Retrieved from: https://e-revistas.uc3m.es/index.php/REVHISTO/article/view/2756

García-Álvarez J, Puente-Lozano P (2017a) A controversial natural border. The making of the Spanish-Portuguese boundary along the Minho River (1855–1866). Geoforum 79:5–16. https://doi.org/10.1016/j.geoforum.2016.11.015

García-Álvarez J, Puente-Lozano P (2017b) Bridging central state and local communities' territorial visions: boundary commissions and the making of Iberian borders, 1750–1900. J Hist Geogr 57:52–61. https://doi.org/10.1016/j.jhg.2017.05.007

García-Álvarez J, Puyo J-Y (2015) La aportación geográfica y cartográfica de las Comisiones de Límites luso-franco-españolas (ss. XVIII-XIX). Apuntes para una historia comparada del conocimiento y la cooperación científica sobre la frontera. Terra Brasilis (Nova Série), [online], 6. Retrieved from: http://terrabrasilis.revues.org/1693

García-Mañá LM (2005) Couto Mixto. Unha república esquecida. Xerais, Vigo.

Godinho P (2011) Oír o galo cantar dúas veces. Diputación Provincial, Ourense

Herzog T (2015) Frontiers of possession. Spain and Portugal in Europe and the Americas. Harvard University Press, Cambridge

Instrucciones para ejecutar la demarcación de la frontera entre España y Portugal (1895). Imprensa de Libano da Silva, Lisboa

Lobo A (1853) Noções gerais sobre o direito das gentes. Revista Universal, Lisboa

Paiva V (1850) Elementos de direito das gentes. Imprensa da Universidade, Coimbra

Popescu G (2012) Bordering and bordering in the twenty-first century. Rowman & Littlefield, London

Riquelme A (1849) Elementos de derecho público internacional. Imprenta de D. Santiago Saunaque, Madrid

Rueda G, Espinha da Silveira L (1993) Dos experiencias: España y Portugal. Ayer, 9:19–27. Retrieved from: http://revistaayer.com/sites/default/files/articulos/9-1-ayer9_DesamortizacionPeninsulaIberica_Rueda.pdf

Solans B et al (2019) Los Estados nacionales. Emse Edapp, Barcelona

Storey D (2001) Territory. The claiming of space. Pearson, Harlow

Teschke B (2006) The metamorphoses of European territoriality. a historical reconstruction. In: Burguess M, Vollaard H (eds) State territoriality and European integration. Routledge, London, pp 37–67

United Nations (1991) Treaty Series. Treaties and international agreements registered or filed and recorded with the Secretariat of the United Nations, vol. 1288. United Nations, New York. Retrieved from: https://treaties.un.org/doc/Publication/UNTS/Volume%201288/v1288.pdf

Modern Delimitation of the Spanish-Portuguese Border in an Area Subject to a Centuries-Old Dispute: Dehesa de la Contienda (1893–1894)

3

Cristina Santos-Sánchez

Abstract

This work aims to analyse the modern delimitation of the Spanish-Portuguese Border in the Dehesa de la Contienda, an area subject to a centuries-old dispute which was not solved until a specific border agreement signed in 1893. Since the thirteenth century, this territory located between the border of Spain and Portugal was disputed by the Spanish towns of Aroche and Encinasola and the Portuguese town of Moura. In 1542, the three localities concluded a first agreement, called "La Concordata", in order to share the use of this area. However, until the aforementioned agreement the area was neither divided nor demarcated. In order to understand the border dispute of La Contienda and its final resolution, this chapter develops two key points: firstly, the acquisition of geographical knowledge necessary to inform about the conflict, and secondly, the cartographic representations that the Joint Boundary Commission developed to delimit this part of the border. This chapter also addresses the reasons, issues, controversies and final solutions adopted in this area of the border to further understanding of the Commission's phases and conclusions.

3.1 Introduction

Stability of the frontier between Spain and Portugal dates back to the twelfth century. Nevertheless, delimitation did not occur until the nineteenth century, when it became the objective of nation states to discover and demarcate the whole of their territory (Cordero 1960). Despite its stability, the borderline has undergone several variations over the centuries, and a numerous succession of treaties and delimitation agreements have existed between the two countries (García 1988). The treaties of Badajoz (1267) and Alcañices (1297) defined most of this border, known as "the Stripe" (*la Raya* in Spanish, *A Raia* in Portuguese). It is these treaties that make the Spanish-Portuguese border one of the oldest and most stable in the world (Trillo and Paül 2014). However, it was not until the signing of the 1864 *Tratado de Límites* (Boundary Treaty), and the 1926 *Convenio de Límites* (Convention of Limits) that it was legally defined and delimited.

This work has been supported by the Spanish Ministry of Economy, Industry and Competitiveness and the European Regional Development Fund (Research Project CSO2015-65301-P), as well as by the Spanish State Research Agency (Research Project PID2020-114088GB-I00/AEI/10.13039/501100010033).

C. Santos-Sánchez (✉)
Carlos III University of Madrid, Getafe, Spain
e-mail: csantos@hum.uc3m.es

J. García-Álvarez and P. Puente-Lozano (eds.), *Beneath the Lines*, Historical Geography and Geosciences, https://doi.org/10.1007/978-3-030-96904-2_3

In the nineteenth century, when modern delimitation of the borders in the Iberian Peninsula began, territorial altercations and conflict arose at some points along the border between the two countries (Barreiros 1961–1965; García 1993; Godinho 2011; Kavanagh 1994). On occasion, these disputes had significant repercussions on delimitation, as was the case of Olivenza, which completely paralysed the process and meant that it was not possible to sign a single treaty that would include the whole border. These territorial conflicts, known as *contiendas* (disputes) in the context of the Spanish-Portuguese border, would be dealt with and solved by Boundaries Commissions and Treaties (García-Álvarez and Puyo 2015). In the case of Spain and Portugal, the majority of these conflicts originated in areas that were the subject of litigation or territory whose sovereignty was not recognised (Herzog 2015).

The first contemporary treaty, the Boundary Treaty signed in Lisbon on 29 September 1864, established the boundaries of the border from the mouth of the Minho River to the confluence of the Caya and Guadiana Rivers (Cairo and Godinho 2013). The second and final treaty, the Convention of Limits, signed on 29 June 1926 in Lisbon, delimited the frontier from the confluence of the Cuncos and Guadiana Rivers to the latter's mouth in the sea. The only part of the border that remained without an agreed delimitation was the area that followed the course of the Guadiana between the Caya and Cuncos Rivers, which coincides with the territory of Olivenza/Olivença. Before the 1926 Convention, Spain and Portugal had signed an agreement in Madrid on 27 March 1893 that definitively fixed the boundaries of the Dehesa de la Contienda, the area that is the subject of this study. Finally, the two countries signed the Demarcation Act relating to La Contienda in Tojal Alto on 18 July 1894. The main aim of the treaties on boundaries, in particular of the two treaties that Spain and Portugal signed, was to reinforce state control over the borderlands, end territorial conflicts between frontier communities and stamp out smuggling (García-Álvarez and Puyo 2015).

Dehesa de la Contienda (Fig. 3.1) is located at the southern end of the Spanish-Portuguese border between the province of Huelva and the region of Alentejo. Its peculiarity stems from the dispute for possession and exploitation of the land, which was under common use by the towns of Aroche, Encinasola and Moura.

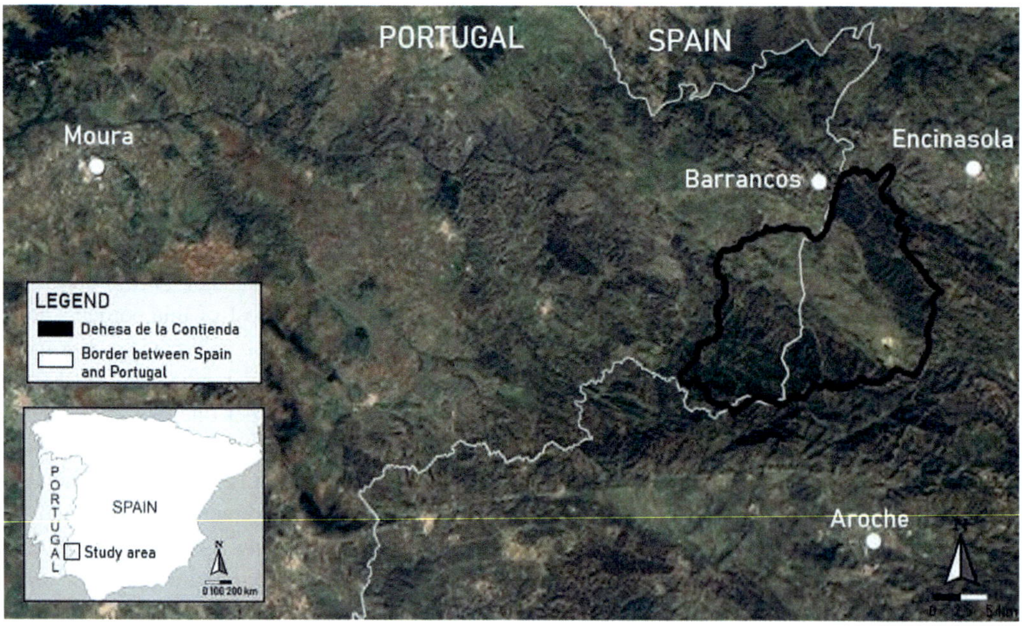

Fig. 3.1 Location of Dehesa de la Contienda. *Source* Prepared by the author

This study highlights the contribution of the Joint Boundary Commission on the stretch of the border known as La Contienda. The role of the Commission was significant for the maps it produced (Dias 2009) in its attempts to solve the conflict. This study consists of three sections and conclusions. The first section deals with the origins and historical context of La Contienda to give an understanding of the magnitude of the conflict and the development of this territory located between Spain and Portugal. It also shows how conflicts have considerable repercussions when it comes to delimiting a border. The second section discusses the unsustainable nature of the conflict in the nineteenth century, the negotiations held throughout that century, the maps drawn, and the various controversies. Finally, the third section presents the result: the final project to divide La Contienda and demarcate the borderline between Spain and Portugal.

The information presented in this study is for the most part a compilation of source materials from the diplomatic and military archives in Lisbon and Madrid that are the repositories for documentation both on the treaties and the Joint Boundary Commission. Among these are the *Archivo Histórico-Diplomático do Ministério dos Negócios Estrangeiros* (Historic Diplomatic Archives of the Portuguese Ministry of Foreign Affairs—AHDMNE); the *Archivo Cartográfico y Estudios Geográficos del Centro Geográfico del Ejército* (Cartographic and Geographical Studies Archive of the Spanish Army Geographic Centre —ACEGCGE); the *Archivo Central del Ministerio de Asuntos Exteriores y Cooperación* (Central Archives of the Spanish Ministry of Cooperation and Foreign Affairs—AMAEC); and the *Archivo Histórico Nacional* (Spanish National Archives—AHN).

3.2 Origins of La Contienda and Brief Historical Context of the Need for Delimitation

The 1864 Boundary Treaty between Spain and Portugal delimited the border from the mouth of the Minho and the confluence of the Caya and Guadiana Rivers. From the time it was signed, the intention was to draw the borderline as far as the mouth of the Guadiana. The idea was to base the border on the line followed by the course of the Guadiana, including the area located between its tributaries, the Cuncos to the north and Chanza to the south. However, in subsequent decades, no agreement was reached on delimitation for the remaining stretch of the border. What might have seemed a simple delimitation encountered several complications that proved difficult to overcome and apparently would remain unsolved, thus leaving full delimitation of the border incomplete. As Francisco Serrat—Chairman of the Spanish section of the International Joint Boundary Commission—summed up in April 1926:

> It beggars belief that over such a long period of time it has proved impossible to reach an agreement. This is not for want of ceaseless negotiations and actions. The case is that the delimitation, which in principle seemed so simple, masks a series of thorny issues that even today, at what is undoubtedly a propitious moment to make great advances in this work, cannot go ahead without uncovering the most serious issues that completely obstruct an agreement.
>
> First are the local interests on both sides, which have constantly been the object of temporary arrangements between the town councils or even between neighbours. Generally, these issues are subordinate to the question of ownership or private possession of the border stripe, and therefore as variable as the arrangements themselves. With such rights based on imaginary or dubious deeds of ownership, if not clear acts of arbitrary appropriation, the International Commission would find no small difficulty in determining the exact line of the border.
>
> On the other hand, these circumstances should not have more importance than they generally do. However, even ignoring secondary interests, there are certain cases where the Commission found itself forced to acknowledge the existence of traditionally recognised areas of land that are clearly subject to litigation and have their own boundaries, known as Contiendas (AMAEC, R15874bis, Exp. 25).

This is the context of the main conflict that this study deals with. As already mentioned, although at first glance it appeared to be a trivial issue, in terms of border demarcation, the opposite proved to be the case. There were three disputes, *contiendas*, in the southern area between Spain and Portugal: the Contienda de Villanueva, the Contienda de Mombuey and, finally, the Contienda de Moura, variously known as La Contienda (The Dispute), the Contienda de Aroche or the Dehesa de la Contienda. In addition to these, there was another, less important dispute, the Contienda de la Galiana, which did not cause great disagreement or gain importance until a later date. There were no real deeds of ownership or formal sovereignty, and the inhabitants on both sides of the border used the areas simultaneously or alternately. Most of the areas subject to litigation were under cultivation or used as pasture. Although the answer to some of these disputes resided in an equal division of the land, for instance in the Contienda de Mombuey, it was not possible to extrapolate this solution to the other *contiendas*. The real problem arose when the Joint Boundary Commission attempted to delimit the exact borderline (García-Álvarez and Puente-Lozano 2017). La Contienda was the most complex of the disputes and proved to be the most difficult to delimit. Likewise, it had the largest area of the three disputes, making its division even more complex.

Major conflicts had existed between the towns of Moura, Aroche and Encinasola since the thirteenth century over a broad strip of borderland called Tierra de la Contienda (the Land of the Dispute). Contention over control of the use of this area became, on occasion, violent. Thus, *contienda*, dispute, became the name of the area as the main cause of conflict stemmed from attempts to establish the borders of the three towns, and in each town's pretension to gain the maximum possible jurisdiction (Carmona 1998).

To understand the origins and historical development of La Contienda we must look back to the thirteenth century. The organisational structure for the common use of pastureland implemented by the Muslims after their conquest of the territory served to tighten the bonds between the populations on either side of the Guadiana. A fortified settlement (modern-day Aroche) was established on a hill, which thus became the main centre of population in the area. The later Christian Reconquista brought with it confrontation between Castile and Portugal for Aroche, which Alfonso III of Portugal conquered in 1251 along with Aracena. The conflicts in this border area began with the conquest of Aroche. Later problems arose with attempts to establish the boundaries between the towns of Aroche and Moura, which constantly changed jurisdiction throughout the century (Carmona 1998).

We should bear in mind that awareness of the usefulness of delimitation and the initiative to establish separate spaces between kingdoms arose after the Christian conquest. This was related with changes in the perception of power, which not only refers to power over people, but also over land. The land had to be cared for in order to be productive, allow the population to grow and improve its welfare (Martín 2003).

This was the context in which Alfonso X of Castile and Alfonso III of Portugal signed their first agreement in 1252, which established the dividing element between the two kingdoms as the Guadiana from the confluence with the Caya to the river's mouth in the sea. There were, however, contradictory actions: Portugal continued to intervene in areas to the east of the Guadiana and Castile maintained its connection with the Algarve. In 1253, Alfonso X granted the city of Seville a privilege which fixed its lands and jurisdiction, and Aroche became part of the territory of Seville (a situation that continued until the creation of the modern-day province of Huelva, in 1833). However, two years later, the King of Portugal granted Aroche a municipal charter and from that moment, the area definitively became a disputed territory (AHDMNE, S16E77P3).

Finally, the division again became final in the Treaty of Badajoz, signed in 1267 by Alfonso X and Alfonso III. This treaty confirmed delimitation of the border along the course of the Guadiana, the Algarve passed over to the Kingdom of Portugal and to the north, the border ran along the Caya (Arronches and Alegrete became part of Portugal).

However, although this treaty clearly seemed to delimit the border with the river and boundary marks where no watercourse existed, in practice, the agreements were not respected. The apparent solution to this problem was the Treaty of Alcañices, dated 12 September 1297, which definitively handed sovereignty over Aroche and Aracena to Castile in exchange for certain rights in addition to the municipal areas of Campo Mayor, Olivenza and Ouguela. It likewise established the borderlines between each kingdom. Nevertheless, there was an area between the Chanza and Ardila Rivers that was not divided. This meant that the land of La Contienda remained undivided (Carmona 1998).

Time passed, and it was not until the sixteenth century, faced with an increasingly unsustainable situation, that the kings of Castile (Carlos V) and Portugal (Don Juan III) attempted to intervene in the conflicts that arose in the border area. In 1542, they appointed two plenipotentiaries who, after studying the issue, resolved the conflict. On 21 June 1543, the plenipotentiaries issued their ruling, known as *La Concordata*, on the jurisdiction, borders and use of the pastureland (La Dehesa, until that date known as the Tierra de Contiendas). This agreement between the two countries regulated the rights of the towns of Aroche, Encinasola and Moura. It ordered a division of territory under dispute between the rival communities, but the most controversial area was to be under common use. None of the towns could consider the land as their "own" (Herzog 2015). Exploitation of the Dehesa de la Contienda pastureland was understood to include the conservation and upkeep of the area, grazing for livestock and use of the flora and fauna, which provided the means of subsistence for its inhabitants (Ramos y Orcajo 1891).

Thus, almost 123 square kilometres of land called La Contienda remained under common use. The agreement established landmarks and specified the rules governing management of the area under common use. It also indicated that only the councils of Aroche and Moura would have jurisdiction over the area. The town of Encinasola was excluded from this right, but could exploit the area and give fines in the case of abuses committed, in the same way as the other two towns. This area was for the exclusive use of the inhabitants of the three towns, who could graze their livestock there, and it was therefore forbidden to animals coming from other towns. Likewise, the *Concordata* forbade the use of common land for anything other than animal husbandry. This had previously been a main source of conflict as some inhabitants had attempted to appropriate the land. The zones under cultivation and recognised as private properties were bought up by the three councils and turned over to common pasture. All of the border markers that already existed inside La Contienda were removed, except for those placed by the judges who would annually review the demarcation alongside the mayors of Aroche and Moura (AHDMNE, S16E77P5). From the time of this ruling, the situation remained relatively stable, and although this treaty did not solve the problem, it was the last agreement between the countries before the division of the lands at the end of the nineteenth century. The following section examines the controversies that existed before division of La Contienda.

3.3 Controversy Before the Division of La Contienda

As mentioned above, the basis for delimitation of La Contienda was the signing of the Treaty of Alcañices, when Aroche passed over to Castile and Olivenza went to Portugal. In principle, the justification for this treaty was as an instrument to avoid the conflict that had existed, but its real intent was to share out towns, castles and boundaries. Since all of these are located in the area of the border, we can consider it a borders treaty. However, the word *frontera*, border, does not figure in the text, nor is there mention of any specific element that could establish a borderline: not a river, a hill, or any other object that specifically delimits any point between the kingdoms. Therefore, the treaty is merely a peace treaty, and rather than drawing the borderline, its intent was to solve conflicts and delimit the area by mutual agreement. As it provided no delimitation, no resolution existed and conflicts continued; although "La Concordata" represents a milestone in the history of La Contienda, it did not divide the area that was the cause of most controversy (Carmona 1998).

Since the fifteenth century and for a variety of reasons, Aroche, Moura and Encinasola had been sporadically and continually sacked (Moura against Aroche and Encinasola, and vice versa). In addition to burning the woodland, crops were destroyed and livestock were stolen. The chapter in Herzog (2015) that refers to the dispute traces the origins and historical development of La Contienda, which she describes as a "hydra", in depth. Due to the poverty of the towns, the lack of resources, the lack of fertile land and crops, and the continual theft caused by the fact that there was no border between Spain and Portugal, the towns of Aroche and Encinasola found themselves forced to beg the King for an urgent solution to the situation. Moura did not participate in the attempt to solve the problems, so the Spanish towns assumed that it enjoyed a favourable position. Moura had sufficient farmland and no need to work the land in La Contienda. However, this was not the case because the inhabitants of Moura did farm the land in La Contienda.

From 1750, proposals to divide La Contienda between the three communities became increasingly common. We know that shared use of the area by both the Portuguese and Spanish had been advantageous in the past, but now the opposite was the case. Because, in addition to the existing conflicts between the neighbours, which had repercussions on the relations between the two countries, fertile land that was a highly valued commodity for the Spanish went unfarmed.

After the War of the Oranges ended in 1801, royal officials delimited the border as far as the course of the Guadiana, and they concluded that there was a need to delimit the territory between Spain and Portugal. Two years later, in 1803, a new conflict began between the two countries when Spain again took steps to carry out a division of La Contienda. The main reason behind the need to divide this area was that it was becoming the refuge of delinquents and allowed unhindered passage of contraband (AHN, TR 282, 0431/001). In order to undertake the reform, both monarchs agreed to the formal division of La Dehesa by commissioners from both countries, who they would appoint, in addition to the presence of a representative of each of the towns in the community. The commissioners were Sancho Lopez de Halvra and Diego Rodrigues. Their goal was to mark out the terrain so that it could be used for specific purposes at the same time as serving as a borderline between the two nations (AHDMNE, S16E77P5).

As negotiations progressed, the matter of how to divide the territory caused debate and no agreement was reached. Some of the Spanish proposals were to divide the Contienda into three parts, giving Encinasola, Aroche and Moura the same amount of land. On the other hand, some believed that Moura and Aroche had greater rights than Encinasola, which only had the right to use the land. Conversely, the Portuguese proposals suggested that the division should be into two parts (AHN, TR 282, 0431/001). They alleged that the populations of Aroche and Encinasola together were fewer people than Moura, which had double the livestock. This proposal of the Portuguese Commissioner met with rejection and the Portuguese government informed the Spanish Ambassador in Lisbon that if disagreement continued, the best thing would be to abandon the 1542 agreement and make a division in accordance with "natural" borders (Herzog 2015).

In 1822, new negotiations opened for demarcation of the boundaries and a division of La Contienda between Aroche, Encinasola and Moura, in addition to demarcation of the international border. This round of talks lasted a year and were also unsuccessful (AHDMNE, S16E77P5; González 2018). In 1834, the three towns agreed to introduce regulations that would govern the use of the lands, stipulating which areas each town could use to farm crops. This agreement was ratified on 4 October 1855 (AHN, TR 282, 0431/001). Madoz's geographic dictionary (1849) even describes La Contienda as fertile land belonging to Moura, Encinasola and Aroche, but not to Spain or Portugal. This definition resembles what residents of the three communities believed during the nineteenth century: La Contienda was a neutral territory that was not subject to Spanish or Portuguese law, but had its own mixed legal regime. Later, on 25 August 1863, the commissioners from the three towns drew up regulations for use of La Contienda, which received approval in September that year, and ratification on 14 March 1865 (AHDMNE, S16E77P5). However, yet again, it was not until the end of the nineteenth century that division of the territory took place. The next section explains how the border was finally established and looks at the work the Joint Boundary Commission undertook for the final division of La Contienda.

3.4 Final Process of Dividing La Contienda and the Importance of Its Cartographic Representation

A major figure in the final division of La Contienda was Máximo Ramos y Orcajo, an officer of the Army General Staff (1847–1921), who had been a member of the Joint Boundary Commission since 1882. On joining the Commission, he was a colonel-commandant of the General Staff. He went on to become the Head of the Spanish section of the Commission between 1884 and 1893 and to publish a book on the question in 1891 (Ramos y Orcajo 1891). With the aim of preparing the 1864 treaty, from 1855 on, Spain and Portugal regularly appointed Joint Boundary Commissions, which had the task of preparing and later undertaking the work of border delimitation (García-Álvarez and Puente-Lozano 2015). They were mainly composed of diplomats and the military officials from both countries, and their principal functions were to survey, map and provide a geographic description of the borderlands, draw the border on the ground and erect landmarks. At the time when Ramos y Orcajo joined the Joint Commission, it was conducting demarcation of the border under the terms of the 1864 Treaty on Boundaries, which did not include the sector of La Contienda. However, from 1885, alongside his Portuguese counterpart, Commissioner General Sebastião Lopes de Calheiros, he had the task of reaching agreement on a project for the definitive and equal division of the territory of La Contienda that would go before the two governments. The first task the commissioners undertook was to study all of the background (as explained above) and note any possible errors that might have occurred. The first problem they found was the lack of sufficient information on the map drawn at the start of the nineteenth century. This map (Fig. 3.2), dated 10 October 1803, was based on the survey carried out for the new border demarcation between the two kingdoms, executed by Royal Order dated 14 June 1803 (AHN, TR 282, nº 0431/001).

After having studied this map, they decided that it would be necessary to draw a new, larger scale map and to mark all of the relevant information in detail, thus obtaining an exact idea of the land that needed dividing. This map (Fig. 3.3), which had a scale of 1:10,000 and was reduced to a scale of 1:30,000, served to draw all of the possible divisions in addition to both the Spanish and Portuguese proposals.

This illustrates the importance of cartography in the process of negotiations. To draw up the map, the Joint Boundary Commission conducted two campaigns: one for triangulation in 1886, and another in 1887 to complete details. After the experience of these two campaigns, they reached the conclusion that their work could not be merely verbal, but should also be written. This is the reason why on 14 July 1887 the Spanish Commissioner, Ramos y Orcajo, suggested that negotiations begin even though the map was unfinished. His proposition was to divide La Contienda into three parts, two of which would be equal, for Moura and Aroche, and a third part for Encinasola that would be calculated according to its activity over the last three and a half centuries (AHDMNE, S16E77P5).

Negotiations ended inconclusively on 23 May 1888 because the Spanish Commissioner received no further responses from his Portuguese counterpart. This was the start of the disagreement between Ramos y Orcajo and Calheiros: in addition to not responding, in 1889, Calheiros sent a report he had written entitled

Fig. 3.2 Dehesa de la Contienda (1803). *Source* Francisco de Fersen (approved in Badajoz, 26 June 1804). Scale 1:36,000. AHDMNE, S16.E76. P8. 90,209

Noticia Sobre la Contienda de Moura (News on the Moura Dispute), signed and dated 1 June (AHDMNE, S16E77P3). This report contained slight variations on another he had previously sent to Ramos y Orcajo on 12 April 1888. In response to this, Ramos y Orcajo received authorisation from the Spanish government to refute Calheiros's report in a pamphlet that would be called *Derechos de las Tres Villas Comuneras* (The Rights of the Three Towns over the Land Under Common Use). One of the outstanding aspects of this pamphlet is how Ramos y Orcajo defends his opinion that Calheiros's proposal of dividing La Contienda in half made no sense. Ramos y Orcajo's argument was that, despite being an international issue, which apparently only required a division into equal parts, one for each country, in this case there was no reason why such a division should be made. He pointed out that "…if this were the case, there would be no international questions because a dispute would be solved simply by dividing the object in half" (Ramos y Orcajo 1891).

Likewise, the most interesting aspect of the pamphlet for this study is the information about how Ramos y Orcajo (1891) mentions the three proposals made by Calheiros for the division. The first two proposals supposedly came from the Spanish government in 1805, and the third, which Calheiros defended as the best solution, was his own. The first proposal (the red line in Fig. 3.3) gave 43% of La Contienda to Portugal. This was the graphic representation of the survey made in 1803, and the solution supposedly offered by the Spanish Ambassador on 19 September 1805. Nevertheless, this proposal did not correspond to what was really suggested due to a mistake in the map; in reality Portugal would have taken 37.5% and Spain 62.5%. This proposal was, therefore, invalid.

The second proposal (the blue line in Fig. 3.3) granted Spain 52% and Portugal 48%.

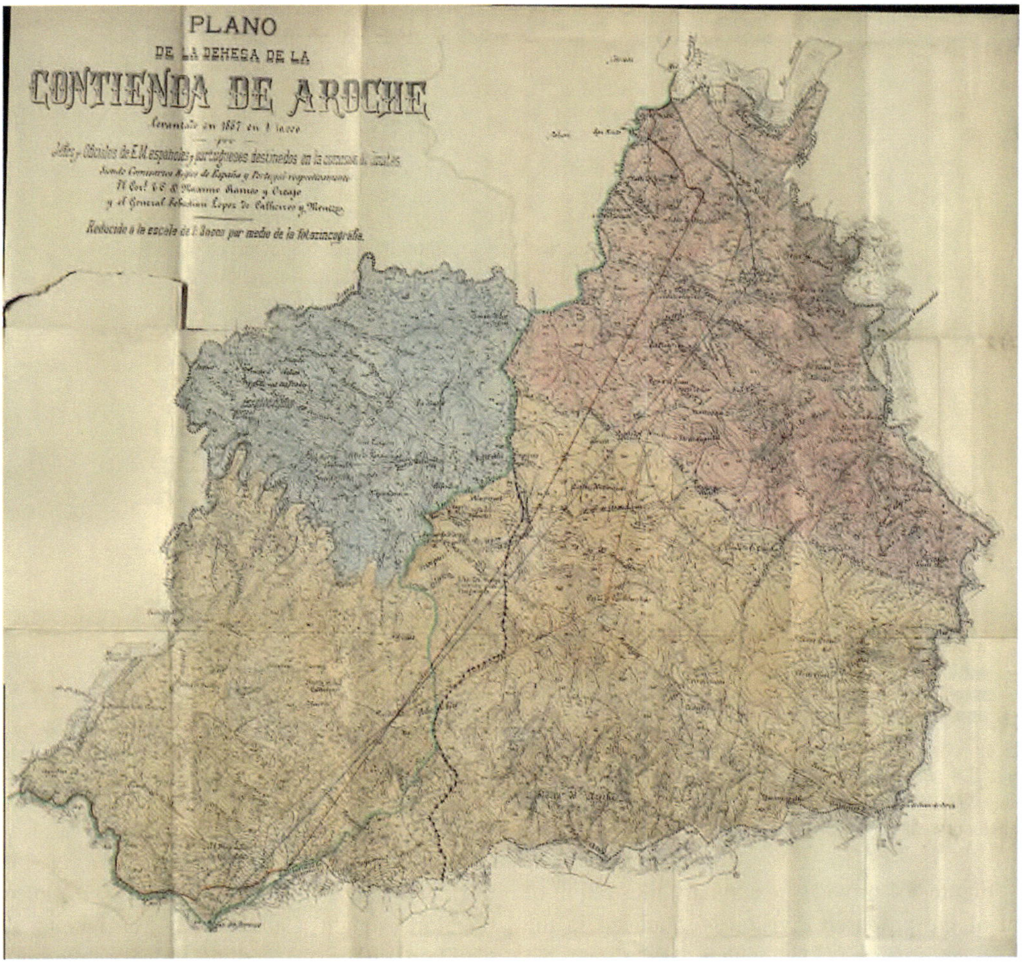

Fig. 3.3 *Plano de la Dehesa de la Contienda de Aroche* (Map of Dehesa de la Contienda of Aroche), 1887. *Source* Comisión de Límites (Boundary Commission), 1887. Scale 1:10,000 and reproduced at a scale of 1:30,000. AHDMNE, S16.E76. P8

This proposal was supposedly made by the Spanish Commissioner in 1805, but this is not believed to be credible, because Portugal would have accepted the division. Moreover, it was based on a map that was found in the file of the negotiations held at that time, and on which two pencil-drawn lines appear. However, there is no indication of whether the lines were drawn by the Spanish or Portuguese plenipotentiary. The third proposal (the orange line) was Calheiros' own. It divided La Contienda in half between both countries. However, it is discussed that this was actually 52% of the area for Portugal, which meant that Spain would receive five square kilometres less. Ramos y Orcajo completely refused this proposal because it favoured Portuguese interests and ignored the reasons why the Spanish claimed the larger part of the territory. Therefore, the proposal that the Spanish Commissioner made was that Portugal should receive 37.5% of the land so that the border would follow the current boundaries between Moura and Encinasola from north to south, and then those of Moura with Aroche for a small stretch, and finally the line of a tributary stream of the Mortigón and a road. The green line in Fig. 3.3 indicates this proposal.

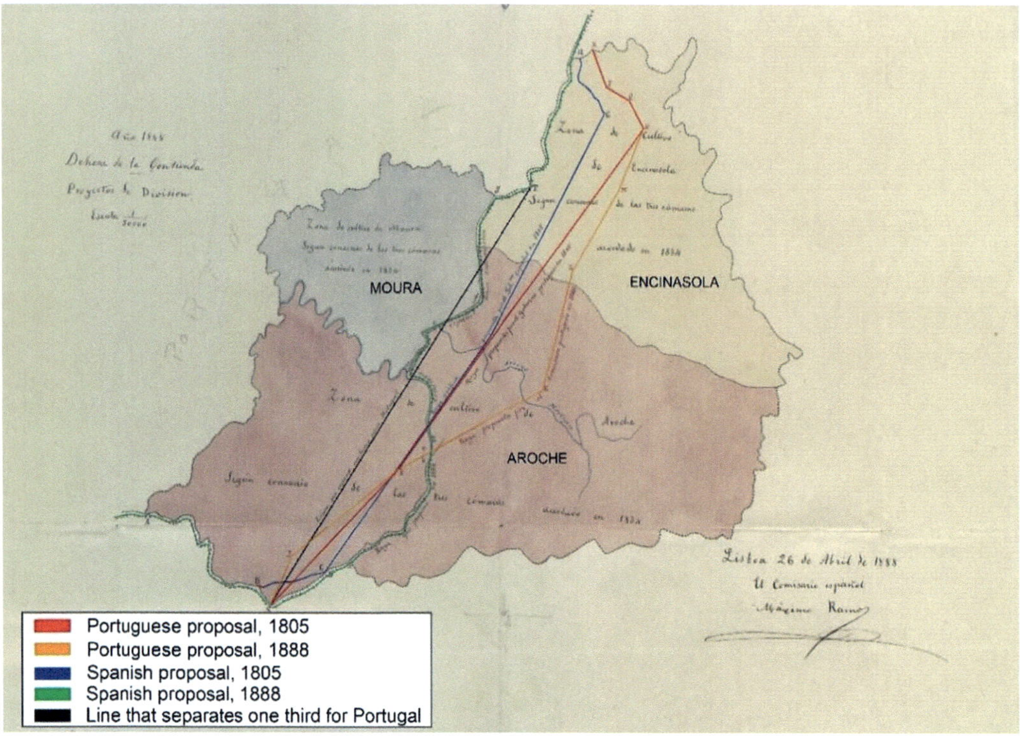

Fig. 3.4 *Proyectos de división de La Contienda en el siglo XIX* (Projects for the division of La Contienda in the nineteenth century). *Source* Máximo Ramos y Orcajo, 26 April 1888. Scale 1:50,000. ACEGCGE, Ar. I-T. 9-C. 4–152 and author's own

Figure 3.4 provides a summary illustration of all of the proposed divisions formulated in the nineteenth century. Although it only shows the areas under cultivation by Moura, Encinasola and Aroche in colour according to the agreement of the three councils reached in 1834, it indicates the lines of the proposals made by both governments in 1805, the 1888 proposals of both Commissioners, and the line that separates a third for Portugal.

Given that this issue was not sustainable, and due to the absence of a response from the Portuguese Commissioner, Ramos y Orcajo was obliged to return to Madrid and personally consult the Spanish government. Subsequently, as negotiations had taken place in Lisbon and all of the documentation was there, Ramos y Orcajo decided to return to Lisbon to continue with the division and resume correspondence with the Portuguese Commissioner in an attempt to persuade him to conclude the process amicably (AHDMNE, S16E77P5). Ramos y Orcajo therefore sought Calheiros out at Quinta de Caridido to suggest a division where Portugal would keep 42% of the territory and the rest would go to Spain. Calheiros refused to accept this. However, this time he said he would make a more favourable offer than the first, in writing. He never made the second offer.

At this impasse in the agreement to divide La Contienda, and with no reply from Calheiros, the Queen Regent wrote to Ramos y Orcajo to express her gratitude for his defence of Spanish interests in the question. Negotiations between Ramos y Orcajo and Calheiros ended in April 1889, when the Spanish government appointed the Minister of Spain in Lisbon, Felipe Méndez de Vigo, the new head of negotiations alongside the Portuguese Minister for Foreign Affairs, Ernesto Rodolpho Hintze Ribeiro. It was decided to move negotiations to Madrid and attempt once more to reach an agreement.

3.5 Final Resolution of the Conflict

When Felipe Méndez de Vigo took charge of talks in 1890, he began preparation of his proposal, ignoring all of the previous negotiations that had not produced results. He only consulted the *Concordata*, which was the only clear and precise text on the question. On this basis, he proposed an equitable solution to Ernesto Rodolpho that, in his opinion, solved the issue and respected existing rights.

Despite Spain being able to claim that the division should be in three equal parts, in the interests of a swift agreement, he proposed another division that was just as fair. His suggestion was to divide the Dehesa in eight equal parts, assigning three to Aroche, three to Moura and two to Encinasola, bearing in mind that the latter had unlimited use of a third of La Contienda. The other possibility was that Moura and Aroche, owners of the Dehesa, should take half of the area each, but compensate Encinasola. The reasoning behind this was the precedent of more than 300 years and the agreement of previous governments from both countries that Encinasola was the recipient of a third of the income. In order to do this, Moura and Aroche would have to cede a third of their part, but as Encinasola did not have any rights of ownership, it could make do with less. Aroche and Moura would have three eighths each, and Encinasola two eighths. What the *Concordata* had stipulated was that division of La Contienda should be into 16 parts, with six each for Aroche and Moura, and four parts for Encinasola.

Ernesto Rodolpho Hintze Ribeiro's replacement as Minister for Foreign Affairs, Ayres de Gouvea, replied to Méndez de Vigo's 1890 proposal in 1892. Ayres de Gouvea was certain that an amicable solution to end this constant international conflict was possible. He proposed that the borderline should be the road that joined Aroche (Spain) and Barrancos (Portugal) that the inhabitants frequently used. With this delimitation decided, on the map Portugal would have 40–45% of the territory and Spain between 55 and 65%. Additionally, Spain would cede part of its territory in order to reach a quick solution. With this resolution, work began on surveying the newly suggested borderlines.

After three years of shared correspondence, an agreement was reached, resulting in the *Convenio de Límites entre España y Portugal en la Dehesa de "La Contienda"* (Convention of Limits between Spain and Portugal in the Dehesa de La Contienda) signed in Madrid on 27 March 1893 (Gaceta de Madrid, 272, dated 29 September 1893). Thus, the first pact for division concluded and the borderline became definitively fixed. This division of La Contienda gave 57.3% to Spain and 42.7% to Portugal. Each portion was subject to the particular sovereignty of the country it was ceded to. The *Acta de Amojonamiento* (Boundaries Act) was signed on 18 July 1894, in accordance with the topographic map of 1887 at a scale of 1: 10,000 (Fig. 3.3).

This clear and concise agreement had four articles. The first article clarified that the division would follow the road used between Aroche (Spain) and Barrancos (Portugal). The second article also specified that the towns of Aroche and Encinasola would maintain full control of part of the land of La Contienda. Encinasola was granted ownership where it had previously only had access and use. Finally, an appendix indicated where the border landmarks should be erected to show the frontier between the two nations. Each marker had a number, a location specified in degrees, the distance to the next marker, and a description of the place of erection, in addition to other observations. The agreement thus provided for 30 landmarks and put an end to years of disputes.

This delimitation came under scrutiny again in the 1926 Treaty on Boundaries (Tratado de Límites, Gaceta de Madrid, 180, dated 29 June 1927) which definitively delimited the Spanish-Portuguese border. It was included in this agreement for reasons of expediency and to complete the text of the Convention of Limits in clear and precise terms, with the addition of certain articles that specifically referred to maps that were mutually agreed by the Commission. This is why Article 7 fully includes the

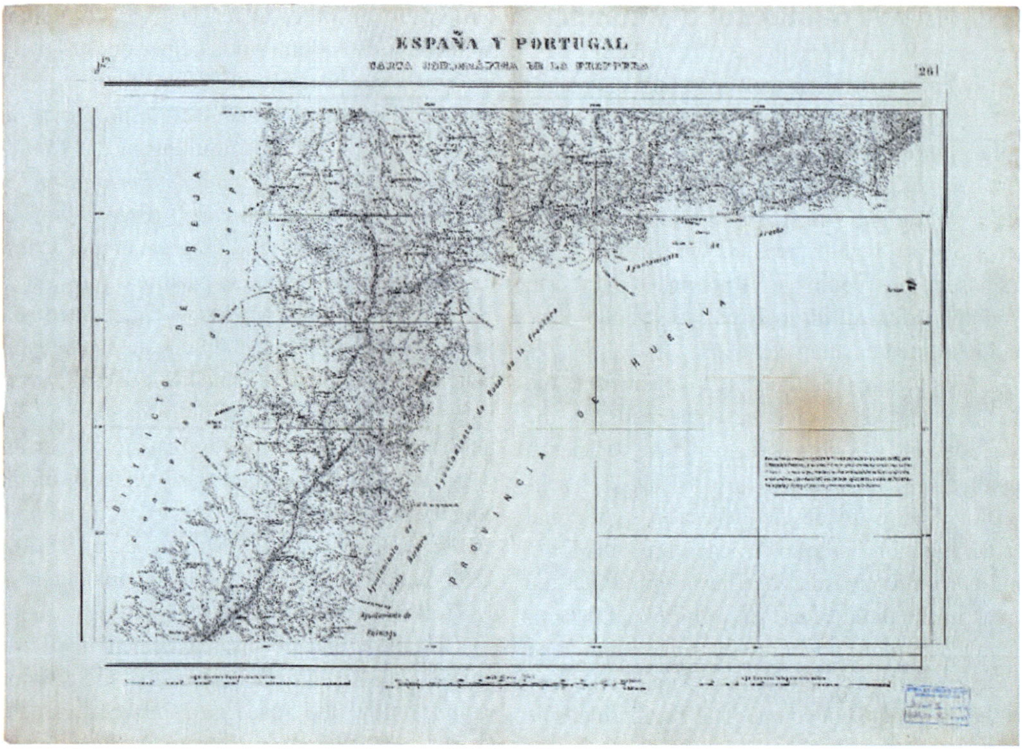

Fig. 3.5 *Carta corográfica de la frontera* (Chorographic map of the border) Indicates the border for the Dehesa de la Contienda established in the 1893 treaty. *Source* Comisión Internacional de Límites entre España y Portugal (International Spanish-Portuguese Boundary Commission) Scale 1:50,000. 1907–1913, ACEGCGE, Ar. I-T. 6-C. 4–36-3

delimitation of La Contienda made in 1893 (AMAEC, R15874bis, Exp. 25). It also includes a topographic map of the sector that shows the borderline exactly (Figs. 3.5 and 3.6).

The fixing of limits and erection of markers for the municipal areas of Aroche and Encinasola went ahead on 7 November 1898. Aroche and Encinasola thus became separate, but they continued to manage the Dehesa de la Contienda, albeit independently. In 1899, demarcation of the two municipal areas finished and the *Concordata* was adhered to until 1 May of the same year, when the towns of Aroche and Encinasola, which were in the Spanish zone, agreed on the division of the area and therefore granted approximately 3000 hectares to each town (González 2018). After several disagreements about this division, the boundaries were finally set. In 1932, the Instituto Geográfico y Estadístico (National Institute of Geography and Statistics) completed the division when the jurisdictions of Aroche and Encinasola finally separated, to achieve final delimitation of the territory. In 1955, the border markers between Spain and Portugal were erected, thus ending a conflict that had lasted more than 650 years (González 2018).

3.6 Conclusions

Delimitation of the rest of the border after the 1864 Boundary Treaty was neither as clear nor as easy as had been expected. Although a signed agreement existed and delimitation work was ongoing, it proved difficult to reach agreement. This was also the case in the area called La Contienda, the subject of this study. After several centuries of conflict and litigation about the zone,

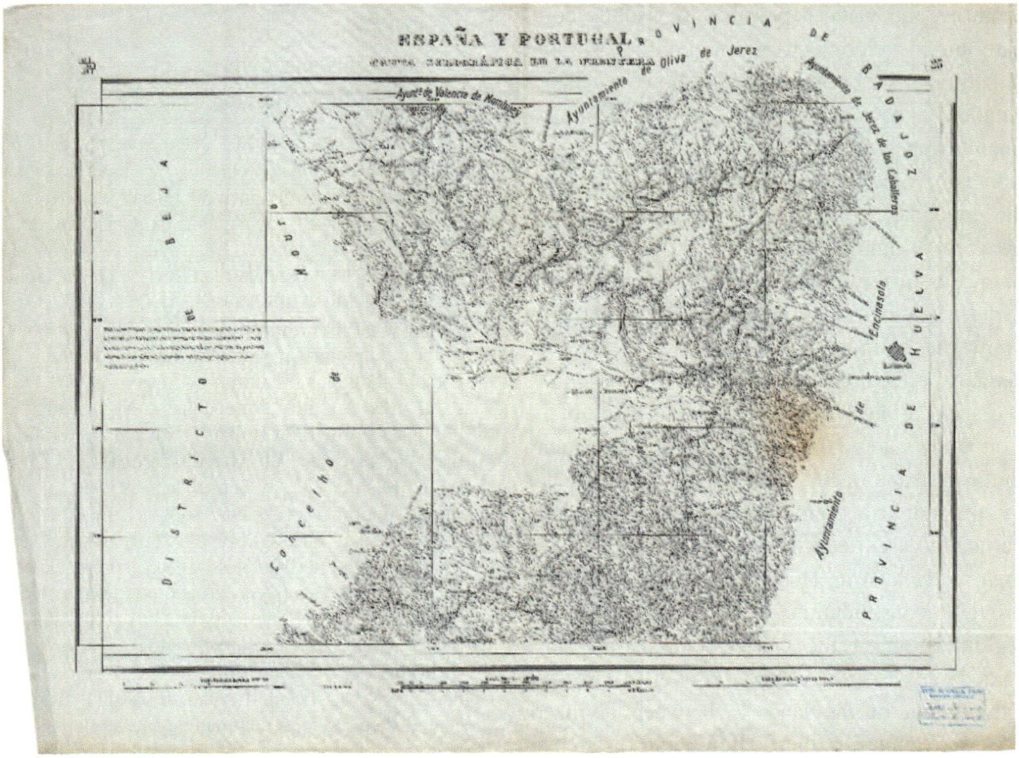

Fig. 3.6 *Carta corográfica de la frontera* (Chorographic map of the border) Indicates the border for the Dehesa de la Contienda established in the 1893 treaty. *Source* Comisión Internacional de Límites entre España y Portugal (International Spanish Portuguese Boundary Commission) Scale 1:50,000. 1907–1913, ACEGCGE, Ar. I-T. 6-C. 4–36-4

the possibilities of the two governments reaching agreement seemed slim. The starting point was when both countries signed the Treaty of Alcañices, which was merely a peace treaty that did not delimit the border, meaning that conflicts continued. The later *Concordata*, rather than doing justice, divided the territory between the rival communities, reserving the most controversial area for common use. Both monarchs maintained their dominions because the question of whether the territory was Spanish or Portuguese was not dealt with. Throughout the nineteenth century, the residents of the three communities thought of La Contienda as a neutral territory that was not subject to Spanish or Portuguese law, but had its own mixed legal regime. After many interruptions, negotiations reached a conclusion at the end of the nineteenth century, but the conflicts would not end until the twentieth century. Although Spain and Portugal agreed the division of La Contienda in 1893 and 1894, this partition was included in the 1926 borders treaty, challenges and protests continued. It was not until 1955 that the division of the Spanish part of the area concluded, when Aroche and Encinasola fixed their boundaries in the Dehesa.

Due to the complexity of delimitation and demarcation of this area under dispute, officers of the General Staff of both Spain and Portugal drew up an extensive collection of maps and reviewed all of the existing historical documentation. The surveying undertaken for the 1803 and 1887 maps constituted indispensable work to achieve the solution of this conflict. Cartography was an ever-present element in the work and for the proposals. It provided a simple representation of the different proposals that had arisen over the centuries and made it possible to visually compare the development of the conflict without difficulty. It was not until 1887, however, that a detailed and accurate map of La Contienda finally appeared. Consequently, the commissioners became experts on the past and present of the area, its conflicts and the discrepancies between its inhabitants, and were able to communicate this information using maps. Nevertheless, while for the state it was important to know the boundaries of its power, what concerned the local communities were the rights to use and exploit the land. The delimitation of La Contienda marked the end of common ownership in the area and of the traditional use of the lands by the border communities. Even so, in this particular case, delimitation was more beneficial than the lack of it. This was not only because it brought consolidation, acknowledgement and legitimisation of the border, but also because shared use of the area did not allow full exploitation of the land "...between civilized, educated countries there should not exist land that cannot be exploited due to confusion over rights..." (La Época 1891).

The real conflict in this area was that after many centuries the towns wished to fix their boundaries and aspired to have the largest possible portion of the territory. For five centuries, 123 square kilometres remained under common use: La Contienda. It was a no man's land and the land of all, until its final demarcation and delimitation. And, as the local inhabitants used to say, "*Las cuentas de La Contienda, que el diablo las entienda*" (Only the Devil understands the accounts of La Contienda).

Archival Sources

ACEGCGE. Archivo Cartográfico y de Estudios Geográficos del Centro Geográfico del Ejército. Madrid.

AHDMNE. Arquivo Histórico-Diplomático do Ministério dos Negócios Estrangeiros. Lisbon.

AHN. Archivo Histórico Nacional. Madrid.

AMAEC. Archivo Central del Ministerio de Asuntos Exteriores y Cooperación. Madrid

References

Barreiros JB (1961–1965) Delimitação da fronteira luso-espanhola. vol. I, 1961. O Distrito de Braga

Cairo H, Godinho P (2013) El Tratado de Lisboa de 1864: la demarcación de la frontera y las identificaciones nacionales. Historia Política 30:23–54

Carmona MA (1998) La explotación Ganadera de la Frontera Luso-española: La Contienda de Moura, Nódar, Aroche y Encinasola. Revista da Faculdade da Letras, Historia 15 Oporto: 241–251

Cordero JM (1960) Fronteras hispánicas: geografía e historia, diplomacia y administración. Instituto de Estudios Políticos, Madrid

Dias MH (2009) Finis Portugalliae - Nos confins de Portugal. Cartografía militar e identidad territorial. Instituto Xeográfico do Exército, Lisboa

Gaceta de Madrid, 272, 29 September 1893

Gaceta de Madrid, 180, 29 June 1927

García-Álvarez J, Puente-Lozano P (2015) Las Comisiones Mixtas de Límites y las representaciones geográficas de la frontera hispano-portuguesa (1855–1906). Revista de Historiografía 23:67–100. https://e-revistas.uc3m.es/index.php/REVHISTO/article/view/2756. Accessed 21 Sept 2020

García-Álvarez J, Puente-Lozano P (2017) Bridging central state and local communities' territorial visions: boundary commissions and the making of Iberian borders, 1750–1900. J Hist Geogr 57:52–61. https://doi.org/10.1016/j.jhg.2017.05.007

García-Álvarez J, Puyo JY (2015) La aportación geográfica y cartográfica de las Comisiones de Límites luso-franco-españolas (siglos XVIII y XIX). Apuntes para una historia comparada del conocimiento y la cooperación científica sobre la frontera. Terra Brasilis (Nova Série) 6. http://terrabrasilis.revues.org/1693. Accessed 21 Sept 2020

Garcia LM (1988) La frontera hispano-lusa en la provincia de Ourense. Boletín Avriense. Anexo II. Museo Arqueolóxico Provincial, Ourense

García LM (1993) Miño: ¿existiu unha fronteira? Galaxia, Vigo

Godinho P (2011) Oír o galo cantar dúas veces. Identificacións locais, culturas das marxes e construción de nación na fronteira entre Portugal e Galicia. Deputacion Provincial de Ourense, Ourense

González E (2018) Los bienes comunales. Evolución y propuestas para el mundo rural. Un estudio de caso: La dehesa de la contienda. Departamento de Estudios Jurídicos del Estado. Facultad de Derecho. Universidad de Alicante/Universitat d'Alacant, Alacant

Herzog T (2015) Frontiers of possession. Spain and Portugal in Europe and the Americas. Harvard University Press, Cambridge

Kavanagh W (1994) Fronteras simbólicas y fronteras reales en los límites entre España y Portugal. In: Donnan H, Wilson TM (eds) Antropología sin fronteras. University Press of America, Laham, pp 75–87

La Época (1891) 29-8-1891, no. 14,007. Madrid

Madoz P (1849) Diccionario geográfico-estadístico-histórico de España y sus posesiones de Ultramar (1846–1850). Tomo XII. Imprenta del Diccionario, Madrid, 304–335

Martín J (2003) La tierra de las "contiendas". Notas sobre la evolución de la raya meridional en la Edad Media, Norba. Revista de Historia 16(1):277–293

Ramos y Orcajo M (1891) Dehesa de la contienda: proyectos de división: origen, historia y estado actual, derechos de Aroche, Encinasola y Moura. Biblioteca de Estudios Arochenos nº1. Asociación Cultural Senebra

Trillo JM, Paül V (2014) The oldest boundary in Europe? A critical approach to the Spanish-Portuguese border. Geopolitics 19(1):161–181

4 From Line on a Map to Realities on the Ground: The French-German Joint Boundary Commission at Work (1871–1877)

Benoit Vaillot

Abstract

This contribution focuses on the French-German Boundary Commission that demarcated the French-German border from 1871 to 1877, after the German annexation of Alsace-Lorraine that followed the French defeat in the War of 1870. The genesis of this border cannot be summarised by the delineation drawn on the small-scale map attached to the Treaty of Frankfurt at the end of the war. It took no less than seven years for actors to draw the borderline in the field. From 1871 to 1877, a Boundary Commission—composed of French and German military and geographic engineers—worked on the ground to make the new border effective. During their activities, they erected more than 5000 border stones, made more than 900 maps and settled numerous conflicts, two of which we focus on. Thanks to the diary of one of the border commissioners and the collections of archives produced by the Commission in France and Germany, we study the birth of a border from a new angle. This study emphasises the room for manoeuvre that these officials had whilst delimiting the border, as well as the role the inhabitants played by means of petitions, letters of protest and counter-mapping. This French-German case will be put into perspective with other Boundary Commissions encountered in our research.

B. Vaillot (✉)
University of Toulouse – Jean Jaurès, Toulouse, France
e-mail: Benoit.Vaillot@eui.eu

B. Vaillot
University of Strasbourg, Strasbourg, France

4.1 Introduction

In September 1870, at the height of the Franco-German war, the geographical and statistical division of the Prussian general staff published a map of the "Territory of the General Government in Alsace".[1] After the French defeat, a sinuous green line was added to this document on a 1:300,000 scale, delineating German territorial claims on paper for the first time (Delahache 1909). The map was attached to the preliminary peace treaty of 26 February 1871: the green line fulfils the function of delimiting the new Franco-German border. The territories ceded by France to the German Empire were not listed, but were vaguely defined as being located east of this border according to the first article. The green line delineated the territorial transfer, but did not specify the territories transferred. It corresponded to administrative boundaries (communes, cantons, districts, departments), as well topographical boundaries (mountain ridges) (Fig. 4.1).

[1]French diplomatic archives (La Courneuve) then AD, TRT 18,710,007. "Das Gebiet des General-Gouvernements im Elsass", lithography by Leopold Kraatz, Berlin.

J. García-Álvarez and P. Puente-Lozano (eds.), *Beneath the Lines*, Historical Geography and Geosciences, https://doi.org/10.1007/978-3-030-96904-2_4

Fig. 4.1 The French-German border (1871). *Source* Prepared by the author

The question of the delimitation of the 1871 French-German border is well known (Hartshorne 1950; Roth 1992, 2008), but without going into detail, it is nevertheless appropriate to explain its two main rationales, which were contradictory.[2] First of all, German nationalists wished to strip all territories from France in which they believed belonged rightfully to Germany, on the grounds that the living populations in these territories adhered to Germanic language and culture, even if they spoke French in certain social contexts and identified themselves as French. This ethno-linguistic argument was supported by Chancellor Bismarck, who sought to achieve German unity under the aegis of Prussia. Thus, the German-speaking areas of Alsace and Lorraine were to join Germany, from which they had been detached in the eighteenth

[2]The question of the French-German border delimitation is detailed in my thesis: The French-German Border (1871–1914). Territories, Sovereignties, Identities.

century. Then, the Prussian general staff sought to secure Germany by ensuring control of the strategic strongholds in Alsace and Lorraine that defended France. However, some of these military places, such as Belfort and Metz, were located in entirely French-speaking areas, where doubts arose as to the ability of the inhabitants to assimilate German culture. The green line drawn on the map synthesised these ethno-linguistic and strategic logics.

Nevertheless, French diplomats managed *in extremis* to negotiate a modification of its route around Belfort, within a radius that had to be determined during the discussions for a final peace treaty. In exchange, only three small French villages, which had become German memory spaces because of the exceptionally violent fighting that took place there during the war, were requested. But against all expectations, at the treaty of Frankfurt of 10 May 1871, a new counterpart was demanded from German diplomats in order to determine the Belfort radius conserved in France. German diplomats, having become aware of the presence of iron in the subsoil of the French villages immediately adjacent to the border, acquired some of them, and used the discussions on the Belfort radius for this purpose. The delimitation of the Franco-German border also followed an economic logic, which explains to a large extent the length of the negotiations. Moreover, a surprising gentleman's agreement between a French diplomat and Chancellor Bismarck led to a few villages enclosed by the green line escaping annexation.

The creation of the border that separated France and the German Empire between 1871 and 1918 is often summed up in this dimension of delimitation alone. More generally, historiography has for a long time limited the study of the creation of borders to the aspects of delimitation only, paying little attention to the aspects of demarcation. It is true that the term "delimitation" tends to absorb the term "demarcation", but these are two distinctive activities—admittedly they are linked to one another.[3] Indeed, public international law distinguishes between the deed enshrined by an inter-state agreement—delimitation—and the deed materialising the border—demarcation.[4] The first one aims to set the boundaries of an area and the latter aims to materialise its layout. It is thus both a different modality of action and also a change of scale.

The materialisation of the border is precisely what is of interest here, when it ceased being a cartographic and textual abstraction. The treaty of Frankfurt stipulated that a "commission internationale de délimitation" in French (*Grenzregulirungs Kommission* in German) to undertake a field visit in order to "lay out the new border". In Europe, such commissions existed at least since the Carolingian era[5] and had increased from the eighteenth century onwards (Nordman 1998). As García-Álvarez and Puente-Lozano (2017) demonstrated, joint boundary commissions were a crucial means of territorialisation within the larger process of state-building in Europe: they made tangible the abstract "border on paper". In the last thirty years, recent studies have restored the activities of such commissions not only in Europe, but in America (St John 2012), in Africa (Lefebvre 2015; Blais 2014), and even in Asia (Ateş 2013) adopting a *from below* perspective.[6]

In the case of the French-German border, this perspective is made possible with the help of original sources: documents from the joint boundary commission on the one hand, and a commissioner's memoirs on the other. The former represents a disparate amount of documents (letters, minutes, reports, registers with descriptions, maps, etc.), scattered in local and national

[3]The phases of allocation, delimitation and demarcation can be identified in Jones and Boggs (1945).

[4]Demarcation work for the 1871 French-German border was handled by a "commission de délimitation".

[5]Indeed, the Verdun treaty of 843 sanctioning the split of the Carolingian Empire included delegates on the field in charge of the delimitation of the boundaries of each kingdom. See Dion (1947), Foucher (1988) or Mieck (1990).

[6]Also, concerning Eastern Asia, I found in the Hokkaidô prefecture archives (Sapporo) in Japan, the documents of the Russian-Japanese joint boundary commission which on the Sakhaline Island demarcated the border drawn by the treaty of Portsmouth of 5 September 1905.

archives, in both France and Germany.[7] The thousands of documents and maps which make up this corpus can seem off-putting. The same is not true for the recollections of Aimé Laussedat, a military engineer who took part in the demarcation process (Laussedat 1902). Although this source is French leaning and contains errors, his input is exceptional: he is the only one giving us an insider perspective of the activities of the 1871 French-German joint boundary commission in the field.[8]

A comparative reading of such technical documents produced by the commissioners as well as the own memoirs of one of them allows for a tangible understanding of the production of a border.

This paper presents an analysis of the French-German joint boundary commission responsible for the demarcation of the border delimited at the Treaty of Frankfurt, highlighting both the interactions between commissioners and local populations involved in boundary making, and the multi-scalar nature of the process itself.

4.2 Preliminary Works

Soon after the Treaty of Frankfurt, the joint boundary commission began preliminary work on the demarcation. It was composed of technical officials, mostly military, as many French as German, who had received scientific training to perform the geographical and cartographical tasks indispensable for their missions. During the demarcation phase, diplomats faded away, as they were no longer playing the key role that they had during the delimitation phase. Instead commissioners Bouvier, Krafft and Laussedat on the one hand, and Bruce, Herzog and Hauchecorne (then Rhein) on the other lead the operations on the field. These commissioners worked respectively under generals Doutrelaine and Strantz, who were in charge of the liaison with their diplomats and general staff for all conflicts unresolvable on site. French commissioners followed the well-known tradition of *ingénieurs géographes* (Konvitz 1987) whilst their German counterparts were trained in military geography, whilst Hauchecorne brought with him his background as a geologist. He was specifically instructed to ensure that as much iron-rich subsoil as possible was annexed to the German Empire (Beyschlag 1902).

One part of the joint boundary commission was gathered in Metz, and another part in Belfort, both places closed to territories negotiated during Frankfurt discussions (Laussedat 1902, pp. 79–84). There, commissioners perused, studied and reproduced the "cadastral maps" of all municipalities adjacent to the new border (Laussedat 1902, pp. 79–84). Indeed, since medieval times, borders in Europe were traced from municipal boundaries—formally parishes (Dion 1947, pp. 79–84). The territorial integrity of this administrative unit was very rarely questioned, as it represented the unit on which basic tax is calculated (Paasi 1997). Surveyors Laloy and Hufnagel sketched out the land register to be taken on site by the commissioners.

During the summer of 1871, several brigades were spread over different lengths of the border before establishing the demarcation—picketing—with the help of municipal civil servants, road, canal and port engineers and at times forestry officers.[9] Thanks to the cadastral maps, the commissioners supervised the operations with the local mayors and local civil servants who provided information about the boundaries of their municipalities.[10] Commissioner's sought to locate existing boundary markers on the ground

[7]AD, 2Q0 203 à 204. Boundary commission between France and Germany (1870–1878); Bas-Rhin local archives (Strasbourg) then ADBR 1 J. Notizen über die Thätigkeit der internationalen Grenz-Regulierungs-Commission und Verlauf der Grenz-Arbeiten, août 1872, 71 AL 7. Grenzregulierung 1871–1879; German federal archives (Berlin) then BA, R 901/36610. Die Grenzrevision zwischen Elsaß-Lothringen und Frankreich 1872–1879.

[8]AD, 2Q0 203. Composition of the boundary commission between France and Germany, March–May 1871.

[9]AD, 2Q0 203. General Doutrelaine to Ministry of Foreign Affairs, 2 October 1871.

[10]AD, 2Q0 203. General Doutrelaine to Ministry of Foreign Affairs, 26 June 1871; Territoire de Belfort local archives (Belfort) then ATB, 1 M 388. General Doutrelaine to sub-prefect of the Territoire de Belfort, 1 October 1871.

demarcating municipalities, counties or state forests (Laussedat 1902, pp. 79–84). Between Sancy and Lommerange for example, they reused forest border stone dating back to 1784.[11] The correct border was thus the one which used ancient boundaries, legitimised by time.

It was common for commissioners to live together during their operations (Rebert 2001). Wilhelm Hauchecorne therefore proposed to his counterpart Aimé Laussedat to stay together under the same roof, which later refused under the pretext that he preferred travelling with his wife. He felt he could not accept this gesture of friendship because the war had led to the dismemberment of his homeland (Laussedat 1902, pp. 85–104). With these two men, one can already see the future rivalry between the French school of geography and the German one at the end of the nineteenth century, as both would become important geographers. The specificity of this joint boundary commission is that it was set up immediately after the peace treaty, literally the day after, when passions had not subsided. Otherwise, commissioners generally had a preliminary agreement about the exact line of the border to follow, pending later changes to be decided on the ground.

The initial programme aimed to complete the demarcation within one year; although this overlooked that fact that on the ground, the process would spark a growing number of conflicts.[12] The section between Luxembourg and the Moselle River will now be explored as an example. The brigade there was headed by Laussedat and Hauchecorne. Local communities expressed their discontent when witnessing the demarcation operations. Inhabitants of Hussigny, Thil and Villerupt pulled pickets out during the night, compelling the commissioners to redo the surveys and peg these again (Laussedat 1902, pp. 85–104). This behaviour was encountered by every joint boundary commission. But here, the inhabitants' patriotism expressed in the French and German archives is not only instrumental. Populations divided by the border are well known by historiography as having been integrated early on into the national "imagined community" in France, despite cultural singularities, mainly due to the early construction of the French state (Maissen 2020).[13] The mayor of Boulange ostensibly took his time in order to hinder the operations. The elected official stated that he was "not in a hurry to become Prussian" (Laussedat 1902, pp. 85–104). Elsewhere, between Briey and Mars-la-Tour, Laussedat was stopped by a woman accusing him of giving up on them, adding: "We will never be Prussians"; which did not fail to annoy Hauchecorne (Laussedat 1902, pp. 85–104).

On another note, the brigade disagreed on the boundary line near Crusnes, which was claimed by both commissioners. The treaty stipulated that the new border must hug the western boundary of Aumetz. Equipped with cadastral maps, Hauchecorne argued that the village of Crusnes belonged to Aumetz, and thus assigned to the German Empire. He wanted this village at all costs, because its soil is ferrous. Laussedat asserted that the village became a municipality in 1833, according to the *Dictionary of municipalities in France*, and was thus not assigned. As one can see, during the 1870s, commissioners could rely on obsolete maps from the beginning of the nineteenth century and on non-technical books. Without an agreement, the issue was postponed and handed over to the governments. In the end, diplomats decided that the French commissioner was right.

A municipality could be deemed to be either French or German depending on one's perspective. For example, Juvrecourt was located within the district of Château-Salins which was assigned to Germans. Yet, the village was located on the southern side of the mountains separating the Seille and Moncel rivers, which should have

[11] Meurthe-et-Moselle local archives (Nancy) then ADMM, 4 M 175. Report from the general forest officer, 5 January 1874, Summary note of the boundary commission.

[12] AD, 2Q0 203. General Doutrelaine to Ministry of Foreign Affairs, 26 June 1871.

[13] This is confirmed by the massive emigration of Alsace-Lorraine inhabitants at the beginning of the 1870s.

theoretically belonged to the French.[14] But according to the green line on the map, Juvrecourt should be annexed to the German Empire. Such contradictions were not unusual in treaties, and it was up to the joint boundary commission to find a solution.[15] General Doutrelaine felt that the map was "small-scaled (1:300,000), that the contours did not appear, that the green line was drawn roughly and that the delimitation of this map should not prevail over the one described in the treaty, when conflict arises".[16] There was no objection from the German commissioners, and so the village remained in France. This example speaks to the fact that there was room for manoeuvre, at the will of the commission, away from any government intervention.

The archives reveal the local population's agency during the demarcation phase. For three weeks in June 1871, residents of Chavannes-sur-l'Étang hoped to remain French, relying on maps drawn for representatives throughout the ratification of the Treaty of Frankfurt. On these hand-drawn maps, put together quite hastily to show the land loss for France, the village was not annexed to the German Empire, even though the treaty left no uncertainty on the issue. Inhabitants made claims relating to the village's geography, history and culture, arguing that their municipality was "primarily French in terms of language and way of life", that it was located "entirely in the Rhône basin" and that it had been "part of France for centuries".[17] They mobilised ethno-linguistic, geographical and historical arguments to remain French at all costs. This mobilisation remained unsuccessful: an ill-conceived map gave inhabitants hope they could escape their fate, the joint boundary commission strictly applied the treaty and made them become German.

This example was not isolated: commissioners collected petitions, complaints and grievances.[18] Inhabitants also produced sketches of their own plots, which were hardly useable for the commissioners.[19] However, some people were more influential than others or at least spared no effort to influence the joint boundary commission. Large landowners are amongst those.[20] In Audun-le-Roman, Robert de Wendel, steward of Villerupt's forge and foreman of forges in Hayange, Moyeuvre-Grande and Stirling-Wendel payed a visit to the commissioners himself (Laussedat 1902, pp. 105–145). The wealthy industrialist family was worried that the new border could dismantle its empire. They submitted corrections for the boundary line: German commissioners were willing to consider them but demanded compensation, which was refused by the French commissioners. Local populations were convinced that the Wendel family aroused the cupidity of the Germans commissioners to move the line westward, rather than eastward.[21]

In late summer 1871, the picketing process was completed along the full-length of the border, but demarcation operations fell behind due to conflicts having emerged about the municipalities of Avricourt, Raon-lès-Leau and Raon-sur-Plaine. By means of prefects, commissioners requested feedback and comments from all corporate bodies about their preliminary works, which would be examined in the autumn. Far from being forgotten, local actors were indeed essential for a clear demarcation of the border. Mayors, local leading figures, took integral part in the activities of the commission. Procurators, rectors and education officers, chief municipal officials, forest officers, presidents of chambers of commerce, bishops, etc. submitted various pieces of information in the best interest of their

[14]AD, 2Q0 203. Minister of the Interior to Minister of Foreign Affairs, 5 June 1871.

[15]The same contradiction appeared between the small-scale maps and the description of the boundary in the treaty of Guadalupe Hidalgo of 1848. See St John (2012).

[16]AD, 2Q0 203. General Doutrelaine to Ministry of Foreign Affairs, 22 June 1871.

[17]AD, 2Q0 203. Minutes from the city council of Chavannes-sur-l'Étang, 7 June 1871.

[18]ADMM, 4 M 175. General Doutrelaine to Prefect of Meurthe-et-Moselle, 2 November 1871.

[19]For the maps produced by the residents reused by the cartographers, see Basset (1998) and Byrnes (2001).

[20]AD, 2Q0 203. General Doutrelaine to Ministry of Foreign Affairs, 1 July 1871.

[21]AD, 2Q0 203. Ministry of Foreign Affairs to General Doutrelaine, 2 October 1871.

respective activities in order to establish a functioning border. Once the preliminary work had been carried out, the installation of the boundary marker was possible.

4.3 Of Border Stones and Maps

In winter 1871, field visits were suspended: as temperatures could be extremely low at this time of the year, particularly along the crest line of the Vosges Mountains.[22] It was the time when commissioners gathered again in Metz in order to plan the final demarcation and to agree on the surveying and description of the new border. For each of the ten sections, a French-German team of two surveyors took over the process, following the work done the previous summer under the supervision of their respective commissioners. The preliminary works were verified according to maps and local residents. Following the laying of boundary markers, each surveyor drew out the delineation of the border and, using a 100-m strip on each side, logged it into a register and recorded it on two types of maps: 1:1250 scale detailed maps giving information on measurements, plots and the location of border stones; and 1:20,000 scale overview maps, made from the land register and enhanced using sketches completed on the ground, which contained the administrative boundaries, the location of official and natural landmarks (and sometimes even the boundaries of banality rights)[23].

Each map had to be certified by the surveyors and then validated by the joint boundary commission (García-Álvarez and Puente-Lozano 2017). Permissible differences could not exceed 1/1000th for the sides of the polygons and 1/500th for angles and apexes. The commissioners relied on measurement instruments invented or fine-tuned in the nineteenth century, such as the aneroid barometer to measure altitude or the theodolite to measure angles. They succeeded in getting a sharper demarcation thanks to geodesy and photography. The making of these maps was then used to produce ones of smaller scales. German commissioners produced more maps than their French counterparts and favoured a larger scale, in order to become better acquainted with local realities which remained foreign to them: 727 sheets at a 1:1250 scale and 22 at a 1:20,000 scale for the Germans versus 101 and 15 for the French, which partly explains the delay with the delimitation. German commissioners were charged with surveying the territory and gathering a wide range of types of information about inhabitants, they took possession of the territory through its systematic cartography, behaviour particularly common in a colonial context (Lefebvre 2015).

The description of the border was achieved with the help of maps, but also recorded in tabular form based on the boundaries of plots on the cadastral maps.[24] Surveyors carefully recorded the position of each boundary marker in registers. The description was both geometrical (angles of the border-polygon, distance between each boundary marker, etc.) and topographical (boundaries of plots, middle point of the ditch, embarkments, etc.). A statement roughly describing the section precedes each register. These registers had to be certified, validated and in line with the maps.

Beyond the abstraction on paper, the border became a material reality, rooted in the land, which remained understudied nonetheless (Thevenon 2005). Since 1872, approximately 5500 boundary markers were placed between Luxembourg and Switzerland by ten brigades from the commission.[25] A total of 4056 major border stones indicating angles were accounted for and another 1500 secondary ones indicating

[22]AD, 2Q0 203. General Doutrelaine to Ministry of Foreign Affairs, 26 June 1871.

[23]ADMM, 1 F 121. Overall plan of the border 1871. ATB SM 98. Overall plan of the border 1871.

[24]AD, 2Q0 203. Schedule for the delimitation, surveying and the description of the international French-German boundary, 29 January 1872; ADBR 1 J. Notizen über die Thätigkeit der internationalen Grenz-Regulierungs-Commission und Verlauf der Grenz-Arbeiten, August 1872.

[25]AD, 2Q0 203. Specifications for the supply and engraving of the border stones intended for the determination of the international French-German border; General Doutrelaine to Ministry of Foreign Affairs, 15 April 1872.

intermediate points, which equals to an average of one every hundred metres, in such a way that one could spot the previous one when located at the foot on one of them. An angle indicating where the border lies was engraved on the top surface of each stone, and its number along with two letters on its visible surface: "F" for the French side, and "D" for the German side. In the course of its work, the joint boundary commission explicitly referred to measures taken by other commissions for other borders, which makes it possible to admit evidence of the circulation of geographical knowledge. For example, the border line was cleared of scrubs for one metre on each side, as the French-Italian joint boundary commission of 1860 had done. The border was thus perfectly visible on its delineation: no one could pretend to overlook it.

The delimitation represented a profitable market for local entrepreneurs, as the operation was estimated to cost 139,000 francs, evenly allocated between the two states.[26] Several entrepreneurs took part in the tender process introduced at the beginning of 1872. Wagner, a German architect recently settled in Metz, won the tender and committed to respect the specification: 26.50 francs per major boundary marker and 22 francs per secondary one. The selected type of stone had to be smooth, uniform, resistant to frost, could not crack nor be produced against the grain. The company worked under the surveyors' scrutiny and was to incur penalties in the event of any delay. Such difficulties did indeed occur in the Vosges region, due to the topography and cold weather; here the installation of boundary markers was only completed in the spring of 1873.

The issue of the maintenance and protection of the border stones rapidly emerged, which the joint boundary commission had not thought about. Some noticeable damages were noticed as early as 1872: the people responsible were almost exclusively residents from the newly annexed territories. Governments encouraged local authorities to "take the necessary measures in order to prevent any wilful or unintended damaging by the nearby populations".[27] The first adopted measure was to make mayors responsible for the damages or destruction of boundary markers on the land of their municipality, in order to stop the "code of silence". A second measure aimed to have repairs or the replacement of border stones be processed by a French-German commission appointed directly by the governments. Each repair or replacement had to obtain the agreement of both parties; minutes had to be ratified by the governments who shared the cost equally. This procedure had to be repeated as many times as necessary and remained in practice until the start of the First World War.

A huge part of the demarcation work was accomplished at the beginning of the 1870s, with commissioners travelling along the border multiple times, picketing and delimiting. At the end of 1873, the border was materialised and tangible. Although most of the work was achieved without any great difficulty, some sections witnessed tensions because of disagreements about the delimitation, which significantly slowed down the work of the commission.

4.4 Two Demarcation Conflicts: The Two Raons and Avricourt

As discussed above, local populations organised themselves to look after their interests. It was not unusual to have multiple interpretations of the border delineation during the demarcation process because the negotiators unknown some realities on the ground (St John 2012). This is what happened in Raon-lès-Leau and Raon-sur-Plaine, where the emerging tensions deserve to be studied in detail. Their allegiance either to France or the German Empire was widely discussed by the commissioners, who also modified the border delineation near Avricourt, due to the location of a railway: indeed, it compromised the establishment of a working border for both the states and residents (Laussedat 1902, pp. 105–145).

[26]AD, 2Q0 203. General Doutrelaine to Ministry of Foreign Affairs, 7 March 1872.

[27]ADMM, 4 M 1716. Minister of the Interior to Prefect of Meurthe-et-Moselle, 15 November 1872.

According to the treaty of Francort, the delimitation line was indeed supposed to be modelled on the western district boundaries of Château-Salins, in the Meurthe department, and the Schirmeck canton, in the Vosges department, where Raon-lès-Leau and Raon-sur-Plaine were located. However, it was also expected that the border followed the ridge line of the mountain which did not follow the municipalities' boundaries exactly, which is why the delimitation line seemed to exclude the two municipalities from the annexation. Due to the contradiction between an administrative logic and a topographical one, the two villages could thus be considered either German or French.[28] Aware of this, inhabitants rallied local and national actors in order to avoid being annexed, even if it cost them dearly.

As the French government claimed to be willing to negotiate, German commissioners demanded the withdrawal of French civil servants from the two municipalities. De facto and *de jure*, the German Empire had sovereignty over Raon-lès-Leau and Raon-sur-Plaine (Laussedat 1902, pp. 105–145). Civil servants were chased out in July of 1871 and the German authorities pressured the municipal council to acknowledge German sovereignty. For example, the French postman collecting the mail was arrested for "theft from the German postal service". Inhabitants submitted a petition to the President, in which they revoked their elected officials considered as traitors. Then they wrote to their representatives in order to put pressure on the government for the villages to remain in France. The situation was so tense that the mayor of Raon-lès-Leau was thrown in jail by the German authorities. The joint boundary commission investigated, and Laussedat concluded that the mayor, "a simple man in both heart and spirit, but also stubborn and an ardent patriot, held out with more obstinacy rather than reason".[29]

The tenacity of the residents paid off as the German commissioners accepted to withdraw the two small villages at the end of 1871, although retaining the state forest and the landlocked properties.[30] Landowners complained, although they were happy not to become Germans. Raon-lès-Leau lost 1100 hectares, which represented nine tenth of its surface area, and Raon-sur-Plaine lost close to 700 hectares, about two third of its surface area. The separated forest areas were added to the German municipality of Gandfontaine, which resulted in a very original demarcation: the new border delineation followed the edge of the forest domains for 18 km, which requires no less than 152 boundary markers. The mayor and priest of Raon-sur-Plaine described the joy of the residents for "being returned to the motherland" (Laussedat 1902, pp. 131–134). However, this joy was short lived, as the village's economy was tied to the forest; residents were dependent on the unrestricted collection of dead wood for cooking and heating, whilst municipalities relied on the forest for revenue.

The French commissioners argued for the government to buy back the forest areas, without success because "Germany considers the mountain and the *col du Donon* to be of the upmost strategic importance".[31] The retaining of the local woodlands allowed the German army to reinforce its strategic position at the mountain top which peaks at 1009 m. Ultimately, the issue was settled: since the "excessive sinuosity of the demarcation line" prevented the effective monitoring of the border by agents from both countries, the commissioners gave back to France in 1872 a forest house and two sawmills, in exchange for the cession of three plots practically hemmed in German territory. Forests were very often the object of conflict during the demarcation phase, due to their strategic importance for

[28]The German commissioners support the connectedness theory (*major pars trahit ad se minorem*): the annexation of the district and the county affects all municipalities within, even the ones located on the western side of the mountain. Laussedat (1902).

[29]AD, 2Q0 203. General Doutrelaine to Ministry of Foreign Affairs, 30 August 1871.

[30]AD, Additional convention to the 12 October 1871 peace treaty, article 10; ADBR 87 AL 581. Die Landesgrenze zwischen Deutschland und Frankreich 1873–1918.

[31]AD, 2Q0 203. General Doutrelaine to Prefect of Meurthe-et-Moselle, 5 September 1872; Moselle local archives (Metz) then ADM 2 AL 3. Abtretung der Gemeinde Raon-les-Leaux.

states and their economic importance for the inhabitants. The French-Spanish boundary commission had also faced this problem in 1847 with the Iraty forest (Puyo 2002). The flexibility of the German commissioners here can easily be explained, they kept everything that had an economic and strategic interest, whilst getting rid of extremely hostile and French-speaking populations.

An issue of a different type arose in Avricourt; however, this time the joint boundary commission immediately reached an agreement regarding the necessary changes. The train station of the village, which was due to become German, was located simultaneously on the Paris-Strasbourg international line and at the junction leading to the small stations of Dieuze in Germany and Cirey in France (Laussedat 1902, pp. 131–134). The commissioners realised that a faithful fulfilment of the border delineation would complicate border controls, railway traffic and the everyday life of the residents all at once. Indeed, the entire municipality of Igney and part of Avricourt would end up hemmed in French territory by the railway line, disconnected from the rest of the annexed territories. It would make it impossible for residents to get to German territory directly without passing through France, unless they cross a level crossing. Another issue was the fact that French trains would have to transit into German territory for a few hundred metres.

French commissioners thus proposed to move the border by 6 kms at the railway line, which resulted in all railway infrastructure remaining in France. This solution solved the problems identified by both parties, but had the disadvantage of divesting the German Empire from having a strategic railway junction and a station located on the international line connecting Paris to Strasbourg, and thus to Berlin. The French proposition was however accepted, on the condition that a new railway station be built on the German side, paid for by the French government.[32] In October 1871, the commissioners sealed the agreement: Igney and the portion of Avricourt located south of the railway line were returned to France. This small plot of land was thus under German sovereignty until 1 January 1872. The portion which remained German, north of the line, took the name of Deutsch-Avricourt.

Whilst awaiting for the Deutsch-Avricourt station to be built, the existing station and the railway infrastructure were returned to France, but the German government enjoyed its use without restriction and governed it in terms of policing, justice and customs. The splitting up of Avricourt had thus resulted in two mirror-municipalities, each being the border stop of their respective state for roads and trains until 1914—a border crossing unintended by the treaty so to speak. In the two Raons, as well as in Avricourt, the integrity of the municipality was not respected in the name of pragmatism and realities on the ground. The settlement of these two demarcation conflicts led to changes in the delimitation line decided by the Treaty of Frankfurt, that is why each one of these joint boundary commission agreements led to the signing of a delimitation agreement by the French and German governments.[33] The demarcation work involved inhabitants mobilisation and resulted in a change in the delimitation of the boundary, reversing the simple centre versus periphery dynamics of imposition and contestation.

4.5 Make Sovereignty Possible

If the demarcation of the border could upset inhabitants, the joint boundary commission could also retain pre-existing relations, or better, make use of their assistance, if this facilitated the exercise of sovereignty, and the work of the customs administrations in particular. The commissioners indeed considered local realities in the minutes of the demarcation in 1877,

[32] ADBR, 1 AL 58. Settlement of the financial dispute between the two municipalities split by the border and the ones restored to France.

[33] Additional Conventions of 12 October and 11 December 1871 to the Peace Treaty.

documentation which marked the end of their activities.[34] This is again quite common to many borders (García-Álvarez and Puente-Lozano 2017), although here they have been reduced to the bare minimum.

The relationship between communities and/or private individuals was affected by the new border. This resulted in multiple transnational landowners with assets scattered in two countries with different legislations and regulations. A harvest coming from a neighbouring field one day was now an imported asset subject to entrance and exit fees. This is why, in order to facilitate "the exploitation of property and forests near the borders", a variety of goods produced within a 10 kms radius on either side of the border became free of customs duties.[35] The cross-border old roads could be used by their owners under the same conditions as if they were entirely located on the same state's territory. These provisions were not unheard of: one finds them in all demarcation operations taking place in Europe at the time (Bouisset 2002). Their concrete application allowed municipalities such as Raon-lès-Leau (see above) to supply sawmills with wood coming from neighbouring German territories.

Commissioners strived to adapt sovereignty in territorial and exclusive terms to all social and cultural realities they encountered. All jurisdictional and ambiguous spaces had to be divided between states so that they became integrated clearly into one or the other (Di Fiore 2017). Article six of the Treaty of Frankfurt had explicitly provided that the diocesan and parish boundaries coincide with the new border. Also, Protestant and Jewish communities that became German were split up from the French consistories. Thus, the demarcation process involved an unexpected actor: the Holy See.[36] A French-German diocesan delimitation commission was formed and implemented the pontifical decrees pursuant to the Treaty of Frankfurt: all the parishes grouping several municipalities crossed by the border were split up.[37] The inhabitants of Ajoncourt and Arraye-et-Han, for example, who were baptised in the same church and who were buried in the same cemetery, were brutally separated by this process. The diocesan delimitation commission followed the demarcation realised by the joint boundary commission.

The demarcation of a border can also be an opportunity to reaffirm existing rights, even ancient ones, between communities. The most well-known historical case is the *lies et passeries* agreement between mountain communities in the Pyrenees concerning forestry and grazing laws on both sides of the French-Spanish border, which for some dates back as far as the twelfth century.[38] The German municipalities of Pettoncourt, Moncourt and Deutsch-Avricourt, for example, demanded and received the right to graze on land located in French border towns.[39] The municipal council of Pettoncourt even unearthed a royal ordinance dated 22 May 1786, logged at the parliament of Metz, to defend this aboriginal right.

When the border cut off a road perpendicularly, the state exerted its sovereignty up to the point where the delineation ran and deployed customs officers with ease. But when the border cut off a road in its length, the collection of customs fees became an issue, as well as the maintenance of the pavement. Indeed, in this case, the exercise of sovereignty was made difficult since residents and customs officers could switch sovereignty as they pleased, by choosing which side of the road to walk on; the maintenance of half of the pavement by each of the states did not make the necessary repairs an easy

[34]ADBR, 87 AL 581. Minutes of the delimitation between France and the German Empire, 26 April 1877.

[35]Additional convention to the 10 May 1871 peace treaty between France and Germany, 11 December 1871, article 12.

[36]Rem in ecclesiastica, pontifical decree of 14 June 1874.

[37]3 ADP 21. Travaux de la commission mixte pour la délimitation des nouvelles circonscriptions diocésaines 1873–1874.

[38]The *lies et passeries* in the Pyrenees were renewed by French-Spanish treaties in the nineteenth century, during the demarcation of the border. See Roige et al. (2002).

[39]AD, 2Q0 204. Minister of the Interior to Minister of Foreign Affairs, 11 May 1875, Minister of Finance to Minister of Foreign Affairs, 5 June 1875.

task. Such a situation, which makes everyday life complicated, pushed the mayors of Moncourt and Coincourt to demand, and later obtain, neutrality for common border roads in terms of customs.[40] The joint boundary commission established a list of 43 roads and common pathways, totaling 43 kms of delineation, split in length for their maintenance and "made neutral in terms of customs"[41]. This meant that residents carrying goods could use the roads without having to pay customs fees, and police officers could travel along these roads uniformed and armed. Each state maintained the stretch of road belonging to the other state, and maintenance costs were equally split.

On multiple sections, the border followed a river bed and thus crossed over bridges, some of which had been destroyed or damaged.[42] Because it was impossible to maintain or rebuild a bridge from one side only, the boundary commission divided the task, with eight bridges for France and nine for the German Empire. Maintenance costs were also split equally, so that none of the states waived their sovereignty over their half of a bridge.[43] Thus, the bridge linking Réchésy to Seppois-le-Bas was entirely maintained by the German administration, but France paid for half of the costs. Furthermore, the commissioners forbid the costs to be paid by a newly introduced toll: residents were thus assured to never have to pay to cross the border via the bridge.

Such adjustments made the life of inhabitants and customs officers easier. The two French municipalities of Vaucourt and Xures, linked by a pathway which went through the German municipality of Lagarde, secured their neutrality, having requested it.[44] For the sake of reciprocity, the same decision was taken for the pathway linking Aboncourt to Bioncourt via the French municipality of Bey. The maintenance of the few metres on foreign soil had to be carried out by the state who made use of it, and maintenance costs were equally split. It was further assumed that local coalitions, led by mayors and councils, aimed to preserve the types of trade and sociability which the border disrupted. For instance, in November 1871 the municipalities of Coincourt, Lagarde, Parray, Vaucourt and Xures expressed their wish to receive similar agreements from the commission, and proposed adjustments to the delineation in order for customs duties to be waived.

The commissioners were particularly focused on forests near the border, due to their economic and strategic importance as illustrated above.[45] Public and private forestry officers received the responsibility of policing the border. Private landowners carried on compensating their own guards, but they had to be nationals and state certified. In France as in Germany, an extraordinary provision with regards to common law enabled the exercising of sovereignty. Local authorities with remaining woodlands on the other side of the border, such as Sancy in the Lommerange area, could request for the neighbouring state to have an officer from the forestry administration police the wooded areas. Public landowners had the privilege to further manage their forest properties according to the rules of the dependent country and to let their employees move freely on both sides of the border.

With the clear distinction between maintenance costs of roads, bridges, etc., the limits of state sovereignty differed from its *exercise*, and that this is a necessary condition for the functioning of a border: the joint boundary commission developed a specific sovereignty regime (Agnew 2005). If the *sovereignty* and its *exercise* would be applied for the exact same territory,

40 AD, 2Q0 204. Minister of the Interior to Minister of Foreign Affairs, 11 May 1875.

41 ADBR, 87 AL 581. Minutes of the delimitation between France and the German Empire, 26 April 1877, article 3.

42 ADM, 15 AL 165. Negotiations with France on the topic of border bridges (1874–1914).

43 ADBR, 87 AL 581. Minutes of the delimitation between France and the German Empire, 26 April 1877, article 7.

44 AD, 2Q0 204. Minister of the Interior to Minister of Foreign Affairs, 11 May 1875.

45 ADBR, 87 AL 581. Minutes of the delimitation between France and the German Empire, 26 April 1877, articles 12 and 13.

with no possible adjustment, residents of Foussemagne would have had to pay customs duties six times when using the road to Chavannes-sur-l'Étang; the French and German states would have had to each build three customs offices, one hundred metres apart, at each point where the road crosses over the border delineation. In order to make the border operational, the customs boundary between the two states was moved to a middle point on the cross-border road: one of the few pragmatic adjustments, aiming to facilitate the daily life of the populations and state action, but which did not change the border demarcation.

4.6 Conclusion

Between 1871 and 1877, the French-German joint boundary commission demarcated the border delimited by the Treaty of Frankfurt. Faced with diplomats' unfamiliar realities, the commissioners roamed the new border laying the border stones, which turned an abstraction on paper into a concrete material border. They even were at the origin of delimitation modifications. Beyond the laying of boundary markers, they also sought to enable state sovereignty by making the border efficient, in order to turn it into a long-term and usable boundary, sparing populations the impracticality of repeated border crossings. The location and reading of original sources allow us to restore inhabitants' agency by looking at documents in which they shared their thoughts with the aim to shed light on how much of a disruption the new border was for them. For this reason, the inhabitants of Raon-lès-Leau were remarkable: they fought at all costs to remain French. This case might seem anecdotal, but the amount of petitions, complaints and grievances in the archives is significant. Besides, these documents were considered, assessed and discussed within the demarcation process of the border. Not only did demarcation agents take part in its delineation and operability, but local populations also played an active role. The decisions and actions of the French-German joint boundary commission challenged the argument that commissioners strictly followed state orders; as García-Álvarez and Puente-Lozano demonstrated for the Iberian peninsula borders, instead these decisions created an institutional space for the circulation of different territorial visions of the border, connecting different actors and scales implied in the demarcation process (García-Álvarez and Puente-Lozano 2017).

At the end of the 1870s, the green line on a map had become a reality. The border was clearly materialised so that no one in France or in Germany could contest it or pretend ignorance of its location. All had been planned, from the maintenance of the boundary markers to the practical exercise of sovereignty. The joint boundary commission operations show us that the production of a border is a long-term, multiscale, political and social process involving multiple actors, and cannot be understood as the drawing of a line on a map. The time elapsed between delimitation and demarcation (six years) was extremely fast in comparison with other European borders. Up until the eve of the First World War, this border was challenged by inhabitants and numerous destructions of boundary markers were recorded. This is why, Aimé Laussedat, wrote at the turn of the century: "this border, well or badly drawn, cannot last" (Laussedat 1902, p. 215). As we know the future proved him right.

Archival Sources

AD French diplomatic archives (La Courneuve).
ADBR Bas-Rhin local archives (Strasbourg).
ADMM Meurthe-et-Moselle local archives (Nancy).
ATB Territoire de Belfort local archives (Belfort).
BA German federal archives (Berlin).

References

Agnew J (2005) Sovereignty regimes: territoriality and state authority in contemporary world politics. Ann Assoc Am Geogr 95(2):437–461

Ateş S (2013) The Ottoman-Iranian borderlands. Making a boundary, 1843–1914. Cambridge University Press, Cambridge

Basset TJ (1998) Indigenous mapmaking in Intertropical Africa. The History of Cartography. University of Chicago Press, Chicago, pp 24–48

Beyschlag F (1902) Gedächtnissrede auf Wilhelm Hauchecorne gelegentlich der Enthüllung seiner Büste in der Aula der Königl. Geologischen Landesanstalt und Bergakademie. A. W. Schade, Berlin

Blais H (2014) Les frontières de l'Algérie coloniale. In: Mirages de la carte. L'invention de l'Algérie coloniale. Fayard, Paris, pp. 197–236

Bouisset C (2002) Le découpage de la frontière franco-italienne dans les Alpes maritimes en 1860 et ses conséquences: entre conflits locaux et enjeux internationaux. In: Desplat C (ed) Frontières. Éditions du CTHS, Paris, pp 167–179

Byrnes G (2001) Boundary makers. Land surveying and the colonisation of New Zealand. Bridget Williams Books, Wellington

Delahache G (1909) La carte au liseré vert. Librairie Hachette et Cie, Paris

Di Fiore L (2017) The production of borders in nineteenth-century Europe: between institutional boundaries and transnational practices of space. Eur Rev Hist 24:36–57

Dion R (1947) Les frontières de la France. Librairie Hachette, Paris

Foucher M (1988) Un modèle géopolitique français. In: Fronts et frontières. Un tour du monde géopolitique. Fayard, Paris, pp 49–72

García-Álvarez J, Puente-Lozano P (2017) Bridging central state and local communities' territorial visions: boundary commissions and the making of Iberian borders. J Hist Geogr 57:1–10

Hartshorne R (1950) The Franco-German boundary of 1871. World Politics 2(2):209–250

Jones SB, Boggs SW (1945) Boundary-making: a handbook for statesmen, treaty editors and boundary commissioners. Carnegie Endowment for International Peace, Washington

Konvitz J (1987) Cartography in France, 1660–1848. Science, Engineering, and Statecraft. University of Chicago Press, Chicago

Laussedat A (1902) La délimitation de la frontière franco-allemande: Souvenirs et impressions. Charles Delagrave, Paris

Lefebvre C (2015) Tracer c'est occuper In: Frontières de sable, frontières de papier. Histoire de territoires et de frontières, du jihad de Sokoto à la colonisation française du Niger, XIXe-XXe siècles. Publications de la Sorbonne, Paris, pp 223–264

Maissen T (2020) Elsass. In: Fahrmeir A (2020) Deutschland: Globalgeschichte einer Nation. C.H. Beck, Munich, pp 199–204

Mieck I (1990) Deutschlands Westgrenze. In: Demandt A (ed) Deutschlands Grenzen in der Geschichte. Beck, München, pp 197–239

Nordman D (1998) L'échange. In: Frontières de France. De l'espace au territoire (XVIe-XIXe siècle). Gallimard, Paris, pp 387–414

Paasi A (1997) Territories, boundaries and consciousness. The changing geographies of the Finnish-Russian border. Wiley, Chichester, New York

Puyo J-Y (2002) Faceries et rectifications frontalières: le cas de la forêt d'Iraty (Pyrénées-Atlantiques). In: Desplat C (ed) Frontières. Éditions du CTHS, Paris, pp 65–78

Rebert P (2001) La Gran Línea: mapping the United States-Mexico Boundary, 1849–1857. University of Texas Press, Austin

Roige X, Ros I, Cots P (2002) De la communauté locale aux relations internationales. Les traités de lies et passeries dans les Pyrénées catalanes. Annales du Midi 114:240:481–499

Roth F (1992) La frontière franco-allemande, 1871–1918. Annales De L'est 1:35–52

Roth F (2008) The making of eastern frontier: the French-German border, 1815–70. In: Altink H, Gemie S (eds) At the border. Margins and peripheries in modern France. University of Wales Press, Cardiff, pp 40–57

St John R (2012) A new map for North America: defining the border. In: Line in the sand: a history of the Western U.S.-Mexico Border. Princeton University Press, Princeton, pp 12–38

Thevenon L (2005) Frontières du comté de Nice. «À la recherche des bornes perdues». Sur l'ancienne limite des royaumes de France et de Piémont-Sardaigne. Serre, Nice

Mexican Geographical Engineers and Demarcation of the Borders of Mexico

5

Luz María Oralia Tamayo

Abstract

Demarcation of Mexico's borders occurred in the nineteenth century. In the field and from offices, Mexican geographer engineers headed the boundary commissions that placed marks on the ground to define the boundaries between Mexico and its neighbouring countries. The northern border, established in the mid-nineteenth century with the United States of America, allowed the young participating engineers to learn about this type of work. Since this border was the result of a war in which Mexico had lost more than half of its territory, the commissioners had to be meticulous and precise, as their counterparts on the US Commission were more numerous and had better instruments. This put the Mexican Commission under immense pressure; however, the commission had an advantage in the excellence of the scientific training the Mexican engineers had received. Demarcation of the southern border with Guatemala and Belize occurred at the end of the same century, and the experience gained on the northern border served as a reference for the work. The engineers in charge of carrying it out confronted many problems during the process, including climatic conditions, with rains and dense vegetation that hindered work, and other drawbacks that the members of Boundary Commission had to solve for themselves. This chapter aims to show how demarcation of the frontiers of Mexico was conducted, who the engineers that served on the commissions were, what difficulties they faced, and what they achieved.

5.1 Introduction

The definition of frontiers is of great importance in the territorial configuration of a country, and in the case of Mexico this was no easy undertaking because the territory underwent a series of changes. During the colonial era, Nueva España (New Spain, present day Mexico) occupied an area of over four million square kilometres. Control of this region was never an easy task for the Spanish Crown because the population was most dense in the centre, whereas the sparsely inhabited north had a mostly nomadic population who did not recognise any authority. To counterbalance this, the Viceroyalty of New Spain established presidios, fortified bases, so that the military sent to the area could protect the missionaries and colonists of these regions. However, the boundaries of New Spain only figured on inexact maps. When the United States of America became independent in 1776, the

L. M. O. Tamayo (✉)
National Autonomous University of Mexico,
Mexico City, Mexico
e-mail: ptamayo@unam.mx

J. García-Álvarez and P. Puente-Lozano (eds.), *Beneath the Lines*, Historical Geography and Geosciences, https://doi.org/10.1007/978-3-030-96904-2_5

Viceroyalty's government became concerned about defining the frontiers of New Spain because US expansion threatened to encroach on the north of the territory. The government of the fledgling country had acquired Florida and Louisiana, but its growing population required new territories. Luis de Onís and John Quincy Adams signed a delimitation treaty defining the frontier in 1819, however, by that date the independence of Mexico (New Spain) was already on the horizon. When independence arrived in 1821, the new Mexican government set up a commission to demarcate the border in accordance with the Adams-Onís treaty. General Manuel Mier y Terán headed the commission, and its other members were José Batres, Constantino Tarnava, José María Sánchez, Luis Berlandier and Rafael Chovel. The group worked between 1821 and 1831 but was unable to complete demarcation (Berlandier and Chovel 1850).

The United States continued to expand, and Anglo-American colonists obtained permits from the Mexican government allowing them to settle in the territories of Texas, which at that time were in the state of Coahuila. The Mexican government undertook to give the colonists land in exchange for their obedience to the government and laws of Mexico. However, the colonists did not respect this condition, and in 1836 the inhabitants of Texas gained independence, which Mexico undertook to recognise as long as there was no annexation to the United States. Again, the condition went ignored, and in 1846, Texas became annexed to the neighbouring country. The boundaries of Texas had not been not demarcated and each country recognised different borderlines; a confrontation between small groups of Mexican and US soldiers gave James Polk, the President of the United States, the pretext to declare war on and invade Mexico. The Treaty of Guadalupe Hidalgo, signed in 1848, ended the conflict, but cost Mexico the loss of over half of its territorial area. In order to demarcate the new border, the governments of both countries appointed a joint boundary commission, whose work began in 1849 and finished in 1857. During this period, Mexico and the US signed the *Treaty of Mesilla* (1853), which modified some of the borderlines fixed in the Treaty of Guadalupe Hidalgo.

Demarcation of the southern border with Guatemala occurred at the end of the same century, and the experience gained on the northern border served as a reference for the work. This border has a different history, related to the 1821 Declaration of Independence. At that time, Agustín de Iturbide wrote the Plan of Iguala, the independence document adopted by most Central American countries that adhered to Mexico, thus also achieving independence. These countries later separated, but a part of Guatemala (Chiapas) remained annexed to Mexico. For this reason, Guatemala had a longstanding claim to the territory, which was finally ceded to Mexico after a popular vote of Chiapas inhabitants to that effect, held in 1824. In 1877, the presidents of Mexico and Guatemala decided to form a Joint Commission to examine and survey the border and propose an agreement that would formalise a boundary treaty between the two countries, which was signed in 1882. Finally, demarcation of the border with Belize went ahead in the last decade of the nineteenth century. Belize passed through several stages: from being part of Yucatan during the colonial period, to a pirate territory, to British Honduras before finally becoming modern-day Belize.

The aim of this study is to show how demarcation of the frontiers of Mexico was conducted, who the engineers that served on the commissions were, what difficulties they faced, and what they achieved. In order to do so, we took the Social History of Science as our theoretical framework and searched for documents in historical archives and map collections. Most of the primary sources were located in the Archivo General de la Nación (General Archive of the Nation/AGN), the Archivo Histórico de la Secretaría de Relaciones Exteriores (Historical Archive of the Secretariat of Foreign Relations/AHSRE) and the Mapoteca Manuel Orozco y Berra (Manuel Orozco y Berra Map Library/MMOB).

This chapter draws also on the works that I have authored or co-authored about the boundaries commissions who worked on Mexico's northern and southern frontiers (Tamayo 2001, 2003, 2004, 2014, 2015; Tamayo and Moncada 2001, 2004; Tamayo and Garza 2007; Tamayo and Amendolla 2016), as well as on other noteworthy studies, such as Paula Rebert's book on the accomplishments of the US and Mexican Boundary Commissions (Rebert 2001), and Nidya Fernanda Caballero's dissertation for her bachelor's in geography on the Mexico-Guatemala Boundary Commission (Caballero 2010).

5.2 Historical Background

The Viceroyalty of New Spain occupied an extensive area. Nevertheless, the northern territories were sparsely populated. "Missions" had been established to take the Catholic religion to colonists and evangelise them, but what the inhabitants referred to as "Indian savages"—due to their lack of fixed settlements—ranged around the area, invading and pillaging villages. Despite the fact that the Viceroyalty's government had established presidios (fortified bases) manned with soldiers who guarded these sparsely populated areas, the vast extent of the territory prevented the soldiers from fulfilling their task in an effective way.

The United States of America had achieved independence in 1776 and European emigrants, who became colonists, demanded lands from what was then a new nation. The US government's aspirations included making this nation a strong, great and powerful country, so the United States began a policy of expansion. The basis for this was the philosophy of Manifest Destiny, which rested on the assumption of a series of "divine rights". Among these, the first was the "right to self-determination", which had allowed them to become independent of England. Equally important was the "right to property" that entitled them to take over land to work on and made expansion an imminent prospect. In turn, the "right to navigation" ensured that the Mississippi river should remain open, while the "right to security" inspired the acquisition of Louisiana and the Floridas by the United States Government. The Napoleonic invasion of Spain facilitated this acquisition as the border had not yet been precisely defined and was, therefore, more vulnerable. In this respect, the government of Spain commissioned Luis de Onís with drawing up a treaty between the United States and Spain. Together with John Quincy Adams, at that time the US Secretary of State, Luis de Onís agreed and signed what became known as the Adams-Onís Treaty in 1819, but as previously mentioned, there was no demarcation of the frontier.

The first decades of independent Mexico were difficult ones: there was a power struggle about which form of government should exist, and more than a decade of war left the population impoverished, hindering proper consolidation of the country. This, to a certain extent, left a situation ripe for invasion by foreign nations.

In 1846, the United States of America was expanding and its government had tried several times to buy territory from Mexico. In the face of the Mexican government's refusal, the United States pursued a tactic of provocation, which was eventually successful when the two armies clashed at the battles of Palo Alto and Resaca de la Palma, also known as Resaca de Guerrero. This gave the President of the United States the excuse to declare war on Mexico. The country was invaded and, despite Mexican defence of its territory, the US army defeated them. The war ended with the signature of the Treaty of Guadalupe Hidalgo on 2 February 1848. Mexico lost more than half of its territory.

After Mexico and the United States signed the treaty, they appointed a commission that would fix the boundaries between the two countries on the ground and subsequently draw the corresponding maps. The young engineers who participated in the commission found themselves obliged to learn about this type of work very quickly. Given that the border was the result of a war, and Mexico had lost territory, the Mexican commissioners had to be meticulous and precise as their counterparts on the United States Commission were more numerous and equipped with better instruments. This put the Mexican commission under a great deal of pressure, and although the Mexican engineers lacked experience in such extreme situations, they did have an excellent scientific training, acquired either at the Colegio de Minería (Royal School of Mines and Mining—an institution founded during the colonial period) or at the Colegio Militar (Military College).

To the south of Mexico, it was a different story. In 1821, to gain freedom from Spain and proclaim independence, parts of what is now Central America participated in the Plan of Iguala, led by Agustín de Iturbide.[1] Subsequently, Central America became separate, but Chiapas (which had been part of Guatemala) continued to remain Mexican in accordance with the wishes of its inhabitants, expressed through a popular vote held in 1824. Guatemala continued to claim this territory and so the definition of the frontier between the two countries was a longer process. By means of an international Convention, in December 1877 the presidents of Mexico and Guatemala decided to form a Joint Exploratory Commission (officially named *Comisión de Reconocimiento de la frontera entre México y Guatemala*) to examine and survey the border, as well as to propose an agreement that would formalise a boundary treaty between the two countries (known as *Tratado Herrera-Mariscal*),[2] which was signed on 27 September 1882 and ratified in May 1883. In order to demarcate the border, the Treaty established a Joint Boundary Commission (*Comisión de Límites*), whose tasks ended in 1902.

In common with the agreement about the northern border, the 1877 Convention and the 1882 Treaty stipulated that both the Exploratory Commission (in the first case) and the Boundary Commission (in the second one) should consist of engineers from the two countries involved, although they could work independently before finally coming together to agree the placement of landmarks on the ground. In the case of the border with Guatemala, demarcation happened at the end of the nineteenth century, and the experience gained on the northern border served as a reference for this type of work. Demarcation of the border with Belize (at that time called British Honduras) did not occur until the last decade of the nineteenth century, as a result of the Boundary Treaty signed by Mexico and the United Kingdom on 8 July 1893 (known as the Mariscal-Spencer Treaty),[3] and completed with four articles added in April 1897. It is worth noting that during the Spanish colonial era, Belize had been part of Yucatán; it subsequently became an area frequented by pirates and, under British rule, was renamed British Honduras (1862–1973), before finally becoming Belize and gaining full independence in 1981.

Three different elements were necessary to establish the frontiers of Mexico with its neighbouring countries. First, the representatives of the contiguous countries had to negotiate and agree to a Delimitation Treaty establishing their frontiers. These treaties required ratification from the respective governments before becoming duly legal. Treaties have a political component that is imprecise and, as would become evident with the progress of demarcation work in the field, it would be necessary to fine tune and later correct details that the treaties could not possibly have taken into account. The appointment of a Boundary Commission, its functions and a provisional date when its work would begin were specified in the treaty.

[1]The Plan of Iguala is a document written to end the interminable war that had exhausted both the country and its inhabitants. Iturbide, who was *criollo* and had been a royalist officer, joined Vicente Guerrero, the insurgent general who fought for independence and an end to Spanish dominion over these lands. Iturbide promoted the agreement throughout the territory and managed to get erstwhile New Spain recognised as a free and sovereign country with the name of Mexico.

[2]Manuel González, the President of Mexico, and Rufino Barrios of Guatemala had appointed Ignacio Mariscal, Secretary of the Office of Foreign Affairs (Mexico) and Manuel Herrera Jr., Envoy Extraordinary and Minister Plenipotentiary (Guatemala), to sign the Delimitation Treaty between the two republics.

[3]The two signatories of this article were Ignacio Mariscal, Secretary of Foreign Relations, on behalf of President Porfirio Díaz of Mexico, and Sir Henry Neville Dering, Baronet of England, Knight of the Most Honourable Order of the Bath, Envoy Extraordinary and Minister Plenipotentiary to the United States of Mexico.

Secondly, the governments appointed a Boundary Commission composed of professionals who, using scientific methods, erected landmarks on the ground to define the borderline. This task essentially depended on the astronomic and topographic knowledge necessary to determine the geographical coordinates of the landmarks. The Commission's work in the field was of great importance, in addition to requiring meticulous care since accurate use of the instruments was necessary in addition to precise calculations of the complicated formulae involved. The task of demarcation was, therefore, essential both to the process of defining the borderline, and to avoid the conflict associated with imprecise interstate boundaries.[4] Thirdly, the engineers on the Commission drew maps of the border that the official representatives of each side would sign to ratify them as official documents. Geographical engineers and topographers were most suited to forming part of this commission and preparing the field plans with the measurements, coordinates and details necessary to draw the final maps.

5.3 The Engineers of the Boundary Commission and Their Scientific Work on the Borders

Given the central role that scientific work occupies in cartography for demarcation and fixing borders, this study focuses on precisely this aspect. The Boundary Commission, in the case of Mexico, was composed of young engineers who had studied at the finest educational institutions of their day, where they had acquired the scientific and technical knowledge they would need. While working in the field, they encountered a variety of problems due to the extreme, difficult conditions prevailing in the region at the time, which ranged from the climate to a lack of resources or hostile human groups. On the northern frontier, this was above all from Apache, Comanche and other groups who opposed demarcation of the territory and prevented the engineers' passage through the region. In the case of the southern frontier, there were loggers who had received, often illegally, concessions to harvest hardwood and were decimating the forests. For all of these reasons, on both the northern and southern borders some commissioners became ill or died[5] in the regions due to illnesses they contracted or injuries they sustained.

Geographical engineers led the commissions, but topographic engineers, both civilian and military, also participated. They had received their scientific training at the Colegio de Minería (Royal School of Mines and Mining) and the Colegio Militar (Military College) and their education had covered knowledge of astronomy, geodesy and cartography among other subjects. Moreover, they knew how to use the precision instruments such as telescopes, theodolites, transit instruments, compasses and chronometers mainly produced by firms like Troughton, French and Roskell (Salazar 1850, p. 15–16). Throughout the process, they transformed information from the instruments to geographic coordinates by means of complicated calculations and with the use of data from astronomical tables and nautical almanacs; finally, points on the line were marked on the ground by erecting the corresponding boundary monument. They defined key points using astronomic methods and determined intermediate points by triangulation, as it was a

[4]Although the task of demarcation has not historically received much attention, numerous territorial conflicts and wars have arisen due to deficiencies in it. For instance, Anderson points out a direct or indirect connection with border disputes in three of every four conflicts in the world, and since the boundaries between sovereign states have great significance in both politics and international relations, maintaining integrity relies heavily on a clear definition of boundaries. This also reduces vulnerability because with recognition of a frontier, there is a line drawn on a map, which can also be identified by means of boundary monuments (Anderson 2003, p. xii–6).

[5]For instance, on the Mexican commission alone, Pedro García Conde died in December 1851 from cholera that he had contracted in the region (AGN Jiménez, 1857: 18–19), Felipe de Iturbide, died from yellow fever in 1853 (AGN, Jiménez, 1857: 97), Luis Díaz had an illness that affected his eyesight and despite treatment he never fully recovered his sight (MMOB, Agustín Díaz 1857, p. 382).

condition that it should be possible to see one point from the previous one.

5.3.1 Analysis of the Northern Border

In an analysis of the northern border, the first noteworthy detail is the extension (3125.3 km) of it. The majority of this boundary follows the course of the Rio Bravo (called Río Grande in the US), which acts as a natural border. From the intersection of the deepest channel of this river with parallel 31° 47′ north, the border proceeds along a line defined by boundary monuments and continues until it reaches the confluence of the Colorado and Gila rivers. From this point, it follows a straight line that divides the Californias until it reaches the Pacific Ocean, to the south of the port of San Diego.

Two treaties determined this borderline. The first, which resulted from a war, was the Treaty of Guadalupe Hidalgo, also called the *Treaty of Peace, Friendship, Limits and Settlement,* signed on 2 February 1848, which granted the United States ownership of what had been more than half of Mexico's territory. John Disturnell published the map that accompanied this treaty, known as Disturnell's Treaty Map, in New York in 1847. The treaty used Juan Pantoja's map, published in Madrid in 1802, to define the location of the port of San Diego. The Treaty of Peace, Friendship, Limits and Settlement of Guadalupe Hidalgo was signed when Manuel de la Peña y Peña was president of Mexico and James Polk was president of the United States, with Luis de la Rosa and James Buchanan as foreign minister and secretary of state, respectively. The authors of this treaty were Bernardo Couto, Miguel Atristain and Luís Gonzaga y Cuevas[6] in representation of Mexico and Nicolas P. Trist as the US representative (AGN, caja 3, exp.1, fi). The second treaty was the *Treaty of Mesilla,* also known as the *Gadsen Purchase*. Signed on 30 December 1853, it modified some of the borderlines fixed in the previous treaty. It was only possible to solve the conflicts that had arisen due to errors in the map used for the first treaty[7] with the sale of another portion of territory.

On the other hand, appointment of the Mexican engineers who formed the first commission took place after the first treaty had been signed. The members of the commission were as follows: General Pedro García Conde (commissioner), who was the commissioner in charge, and the engineers José Salazar Ilarregui (surveyor), Francisco Jiménez and Francisco Martínez de Chavero (engineers first class), Agustín García Conde and Ricardo Ramírez, appointed engineers second class. Felipe de Iturbide was the interpreter and translator.[8] Once the group formed, all of the young engineers travelled to the border in April 1849 by stagecoach, except for Ricardo Ramírez, who had left several weeks earlier with the scientific instruments, luggage and other necessary items. The length of the journey and the bad conditions of the roads affected the instruments, which were already well-worn because of their previous use in classes at both the Colegio de Minería and the Colegio Militar. The engineers met up with the US Commission in order to agree a general plan of operations and begin fieldwork.

The scientific work of the Boundary Commission consisted of two types of task: first, fieldwork, and secondly deskwork.

For the fieldwork, the engineers first reconnoitred the terrain in search of any places with water, set up a camp, installed the apparatus in

[6]Luis Zorrilla points out that Manuel de la Peña, B. Couto, M. Atristain and L. Gonzaga were reviled as traitors by their contemporaries, who did not know the peril that Mexico was in as a nation due to the war and US demands.

[7]Specifically, the map drawn by J. Disturnell included in the Treaty of Guadalupe Hidalgo showed the position of the town of El Paso incorrectly; this detail was a key element to establishing the borderline, which began precisely at this point. The instructions of the governments to their respective commissions caused controversy.

[8]Felipe de Iturbide was the penultimate son of the first emperor of Mexico, Agustín de Iturbide, who had been executed in 1824. The ex-emperor's widow, accompanied by all of her children had taken refuge in the United States, which is why Felipe spoke both English and Spanish (see Tamayo and Amendolla 2016, p.425–447 for further details).

the observatory and selected the terrain on which to base triangulations. Once the instruments were in place, the team conducted astronomical observations that consisted of determining the geographical coordinates of points over the Earth's surface using astronomical methods with the help of astronomical tables showing stellar declinations by date (for the Northern Hemisphere they used the Pole Star, Altair, Alpha Cephei, Arcturus, Rigel, Sirius and others) (Salazar 1850, pp. 51–119). Longitude was with reference to the Greenwich meridian. At the starting point of each stretch, they performed multiple observations, and the coordinates of the point were carved on a plate or in stone on a boundary monument. The third phase of fieldwork was the topographic survey. This consisted of measuring the distances between the points where there were marks, fixing the signs that indicated the distance between them, and rectifying the measurements of the points established using astronomical observations. They also had to indicate geographical features and other details such as ranches, water points, rivers and others on the field plans that would allow the final maps to be drawn. Finally, the culminating moment of fieldwork was placement of the permanent boundary monuments, as these monoliths were the physical indication on the ground of where the borderline passed. Each boundary monument was fixed on the pre-measured mark, and the two commissions agreed the position by consensus; the monoliths were generally cairns made from materials found in the area. However, by the end of the century Apache, Comanche and other groups had destroyed several of these marks, making it necessary to appoint another commission that would rectify and repair the broken boundary monuments or erect new ones.

The second part was deskwork, which consisted of numerical analysis and drawing up the final maps. The engineers spent the period between 1849 and 1855 working in the field, and the cartographic work done between 1856 and 1857 (MMOB, Díaz and Díaz, 1852, 1853–54, 1857).

Demarcation work took place in four stages. The first stage marked out the line that divides the Californias (the peninsula of Baja California that belongs to Mexico, and California, in the United States). This line has two endpoints: one to the west, on the coast of the Pacific Ocean (coordinates latitude 32° 31′ 59″63 north and longitude 117° 08′ 29″7 west). The other is at the confluence of the Gila and Colorado rivers (coordinates latitude 32° 43′ 32″3 north and latitude 114° 36′ 09″9 west) located on the far east of the peninsula. Between these two points, the engineers on the commission determined a further eight points whose coordinates would be specified on corresponding boundary monuments. To determine these points, they used instruments that, although not new, were the best available.

The second stage would begin in the city of Paso del Norte (present day Ciudad Juárez, Chihuahua) and continue until it met the endpoint of the previous stage, in accordance with the Delimitation Treaty. However, when the engineers reached the town of El Paso, problems arose: the treaty signed in 1848 stated that the boundary would be the centre of the deepest channel of the Rio Grande until reaching the town of Paso del Norte. Nevertheless, on arriving at the town, they realised that its location did not coincide with the map used in the treaty,[9] which caused controversy between the commissions of the two countries. The two commissions asked their respective governments for instructions, in the meantime reaching an agreement. They continued marking out the line but a US surveyor, who had been ill for several months and had not participated in the agreement, complained about the line drawn because it stood in the way of a railway project, so it was necessary to negotiate a new treaty to modify this part of the international boundary.

The US commission interrupted their work at this point, and the Mexican commission faced major setbacks when General Pedro García Conde contracted cholera and died in 1852. His son Agustín left the commission, which had to

[9]This part had used the Disturnell map, published in the United States in 1847 (AGN, *Límites enter México y Estados Unidos/Boundaries between Mexico and the United States* caja 3, f.2).

restructure. Another member, Juan B. Espejo, who had only participated for a few months, also left. Salazar, the interim leader of the Mexican commission, continued to work on the survey of the Gila River and of Rio Grande from El Paso to San Ignacio, as stipulated in the 1848 treaty. He also organised the Matamoros Section, who would work from the mouth of the Rio Grande to Laredo. This team consisted of four engineers, two of whom (Francisco Jiménez and Manuel Alemán) were charged with astronomical measurements, two very young topographic engineers (Agustín and Luis Díaz) responsible for the triangulation that would provide details about the course of the river and also for adding the intermediate points between the astronomically fixed ones. In 1853, the section travelled to the mouth of the Rio Grande in the Gulf of Mexico to start work. There, they joined up with the US commission engineers. Their joint task was to locate the deepest channel, travel along the river, astronomically determine the landmarks to define its course, locate the nearby populations and prepare a field plan that traced the course of the river for it to be drawn on the corresponding maps.

The Mexican engineers set up camp at Rancho de La Boca and from a boat located the channel that would indicate the boundary, in keeping with the stipulations of the treaty. According to their agreement with their US counterparts, the Mexican's work would continue as far as Laredo.[10] They marked the starting point of the border between Mexico and the United States in the river mouth. The two engineers who conducted the astronomical measurements placed several landmarks with their corresponding coordinates as far as Matamoros, at which point they moved on to Laredo and Nuevo Laredo to establish the astronomically located points along the river as far as Matamoros. In the meantime, the engineers charged with topography made the triangulations that would provide the detail of the river's physical geography, and placed other intermediate marks that were closer to each other than the astronomically established ones (AGN, Jiménez 1857, pp. 29–75).

The groups made all of the scientific measurements that were subsequently used to draw the maps. The length of the river was such that the commission members of the two countries divided the work in response to pressure from both governments to complete demarcation work, which had already taken longer than envisaged. Due to this pressure, the commissions also divided the cartographic work, with Mexicans drawing some of the maps, while the US commission drew others. An analysis of the maps drawn up by the engineers of the Mexican commission shows that their charts are more detailed than those of the Americans.

At the end of 1853, Salazar returned to Mexico City to participate in discussions aimed at resolving the El Paso problem. The US government wanted to purchase more territory and made several proposals to Mexico. However, the Mexican representatives, among whom was Salazar, did not agree to the sale of more land. Finally, President Santa Anna accepted the proposal that ceded least land, and sold the territory of Mesilla to the USA, putting an end to the problem. On 30 December 1853, the new boundary treaty, known as the Treaty of Mesilla or the Gadsen Purchase, was signed (AGN, *Tratado celebrado con Estados Unidos de América, el 30 de diciembre de 1853.../Treaty signed between the United States of America, 30 December 1853...*). This modified part of the previous treaty and meant that the work on the Gila River was no longer valid.[11]

[10] Post-war Laredo was located north of the river and was therefore in the United States. Popular myth tells how, when the inhabitants of Laredo realised that their town had passed to the US side of the border, the townspeople dug up their dead, crossed the river with them and founded Nuevo Laredo, where they buried their ancestors.

[11] The US government wished to purchase a large stretch of Mexican territory and sent James Gadsden to broker the purchase. Gadsden presented several options, one of which consisted in purchasing Baja California, a large part of Coahuila and almost half of Tamaulipas. Within this context, the sale of Mesilla was the option that ceded least territory.

In 1854, the engineer José Salazar Ilarregui, leading the Mexican commission, requested improved conditions for the group. He received new measuring instruments and more commissioners. The new members were the young engineers Manuel Fernández Leal, Francisco Herrera, Miguel Iglesias, Ignacio Molina and, as attaché, a very young Antonio Contreras, who joined Francisco Jiménez, Manuel Alemán and the brothers Agustín and Luis Díaz.

Thus, the third stage began. Work on the Rio Grande from its mouth to Laredo was about to end and Salazar, with the new members, organised three sections to work in accordance with the new treaty. He formed the Sonora Section with the members of the old Matamoros Section (Jiménez, Alemán and the Díaz brothers). The Sección del Norte (Northern Section) was composed of Salazar, Molina and Contreras. The Rio Bravo Section included Fernández Leal, Iglesias and Herrera (AGN, Jiménez 1857, pp. 112–137).

Salazar trusted in the successful completion of the borderline demarcation. However, despite the promise of more support from Santa Anna's ministers, the country's permanent state of war and a lack of real understanding of the importance of the type of work that the engineers would carry out on the border meant that the resources did not properly arrive. Felipe de Iturbide, who was the legal representative and in charge of securing resources for the commission at customs houses, had died from yellow fever. These circumstances drove Salazar Ilarregui to despair, to such an extent that he complained and demanded resources for the commissioners, but the government did not understand the situation and ordered his detention and imprisonment until he apologised to Santa Anna's government.

With Salazar in prison, Francisco Jiménez reached an agreement with a group of US commission engineers that their Sonora Section would work on marking out the azimuthal line at the intersection of the parallel latitude 31° 20′ north and meridian 111° west as far as the point located in the Colorado River near its confluence with the Gila River. This line was particularly difficult because it passes through the Altar desert in Sonora (Mexico) and the Arizona desert in New Mexico (United States). William Emory, the US commissioner, supported this agreement and recommended that Jiménez direct the joint work of the Mexican and US engineers, and this stretch of the line was duly completed.

The other two sections encountered difficulties due to the lack of leadership, as Salazar was imprisoned and the young engineers that had recently joined the expedition did not have the necessary experience. Work continued until Salazar re-joined the commission. The engineers were able to demarcate parallels latitude 31° 47′ and 31° 20′ north, and meridian longitude 108° 13′ 24″0.10 west, which defined the line between those two parallels. They drew parallel 31° 20′ to the west as far as meridian 111°, where the previously mentioned azimuthal line begins. To conclude fieldwork, they placed monuments (Tamayo 2001, pp. 152–156).

The fourth and final stage consisted of drawing maps based on the field plans in accordance with the points whose coordinates the engineers had determined using data from their astronomical observations and the topographic measurements obtained from triangulation. The maps drawn between 1856 and 1857 are in both Mexican and US archives (Fig. 5.1).

Demarcation work on the northern border was not easy. In most cases, the commissioners had to set up camp in isolated, unpopulated places, and this circumstance left them open to all manner of setbacks, such as the attacks of Apache, Comanche, Coco-Maricopa and similar groups, who stole their food supplies. Although Salazar had requested an escort similar to the one that accompanied the US commission, the few soldiers assigned to protect the Mexicans had neither the necessary training nor discipline since local authorities pressed the into service; the soldiers deserted as soon as the opportunity arose, or were insufficient in number to defend against the tribal attacks.

On the other hand, the commission faced the harsh weather conditions of these regions: at times there was no drinking water to be found in the extreme desert heat, and other regions were so rainy that the land was waterlogged, the rivers

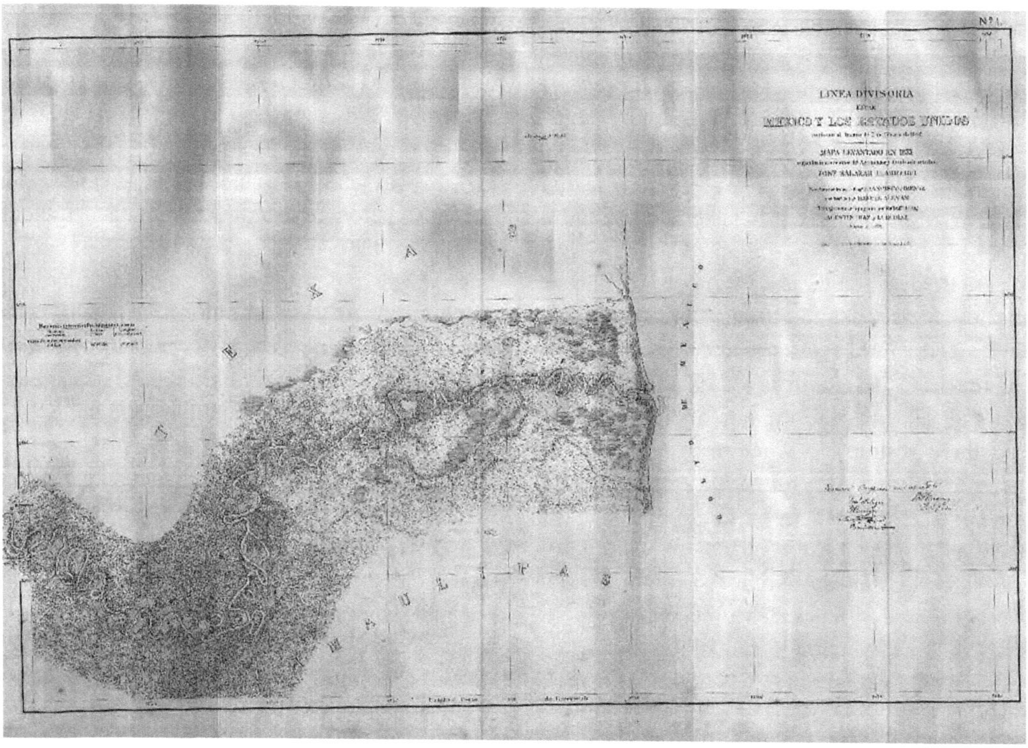

Fig. 5.1 *Comisión de Límites Mexicana, 1857, De la desembocadura del Río Bravo a Matamoros*/Mexican Boundary Commission, 1857, From the Mouth of the Rio Grande to Matamoros. *Source* MMOB

flooded and rain prevented their work. They were also prey to diseases and illness. Apart from Pedro García Conde and Felipe de Iturbide, who lost their lives, Luis Díaz's sight was affected (MMOB, Díaz and Díaz 1853–54, pp. 382–383), and the lack of medical care worsened his physical condition, which he continued to suffer until his death.

In brief, the fieldwork undertaken to demarcate the border consisted in surveying the Rio Grande (Río Bravo) from its mouth in the Gulf of Mexico to parallel latitude 31° 47′ north. At this point, a line was drawn on the ground that followed the parallel until its intersection with meridian longitude 108° 13′ 24″0.10 west, where it changed direction to the south until it met parallel latitude 31° 20′ north. The line follows this parallel until its intersection with meridian longitude 111° west, at which point it begins an azimuthal line to the northeast. It passes through the desert until reaching a point located in the Colorado River, whose course it follows until its confluence with the Gila River, from where it follows a straight line west until terminating in the Pacific Ocean. This line divides the Californias (Fig. 5.2). This stage of work terminated with the placement of boundary monuments on the ground. It is important to indicate that the training and commitment of the Mexican engineers gained the respect and esteem of their US counterparts and, as their diaries show, most of the participating scientists cooperated and worked harmoniously.

5.3.2 Analysis of the Southern Border

As there was no delimitation treaty for this area, it was necessary to undertake the complex preparatory work of exploration of the terrain. This lasted several years and gave time to draw up a proposal for the two delimitation treaties, which were ultimately signed in 1882 with Guatemala and in 1893, with Belize. This border

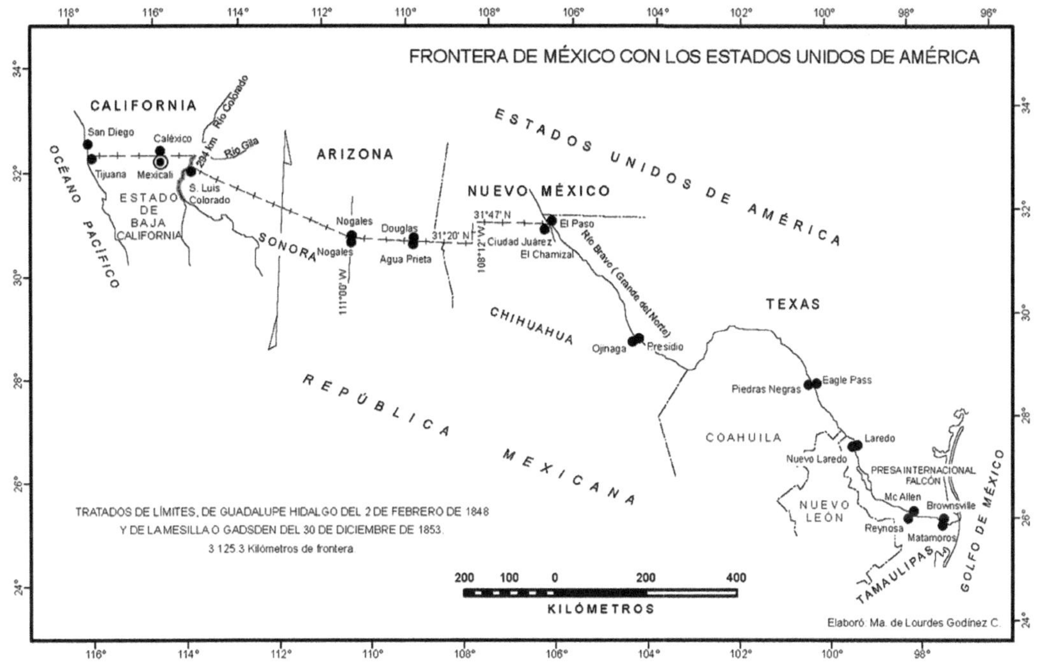

Fig. 5.2 Border between Mexico and the United States. *Source* Tamayo 2014: 146, based on MMOB and AHSRE

has an extension of nearly 1240 km (962 km for the Mexico-Guatemala boundary and another 278 km for the border between Mexico and Belize, including the maritime boundary in the Bay of Chetumal). It is defined by natural boundaries mainly formed by rivers (Suchiate, Usumacinta and Chixoy between Mexico and Guatemala; Azul and Hondo between Mexico and Belize), as well as by lines drawn on the ground. The demarcation lines are defined by parallels 16° 04′ 32″, 17° 15′ 08″ and 17° 49′, and by meridian 90° 52′ 29″, in addition to other natural landmarks such as the Tacaná volcano and Ixbul hill (Pastrana 1914, p. 156).

With regard to the Mexico-Guatemala border delimitation process, it is important to bear in mind that, as indicated above, in 1821, in order to gain independence from Spain, several countries that now form part of Central America joined the Plan of Iguala alongside what was then New Spain. The Central American countries later separated, except for the territories of Chiapas and Soconusco, which would belong to the erstwhile Captaincy of Guatemala. For almost all of the nineteenth century, Guatemala strove to recover these areas, but the inhabitants of Ciudad Real, the capital of Chiapas, voted to remain annexed to Mexico. As indicated above, in 1877 these disputes prompted the then presidents of Guatemala and Mexico to appoint an Exploratory Commission (*Comisión de Reconocimiento de la frontera entre México y Guatemala)*, composed of engineers who represented both countries. Its purpose was to conduct analysis on the ground of the possible places through which the boundary between Mexico and Guatemala would pass. The Exploratory Commission met in Tapachula and signed the protocol that specified how fieldwork would go ahead in 1878. The Mexican Section members were: José Salazar Ilarregui[12] (leader and 1st astronomer), Jacobo Blanco (geographical engineer and 2nd astronomer),

[12]José Salazar Ilarregui had participated in and led the Mexican Boundary Commission for the border with the United States in 1849–1857. However, by the time of this new commission, more than 20 years had elapsed and his physical strength and health were not ideal. He eventually had to be carried down from the peak of the Tacaná volcano on a stretcher due to the emphysema that forced him to abandon the expedition.

Miguel Ponce de León (topographer 1st class), Carlos Revilla, Ignacio Monterde and Manuel Pastrana (engineers), Rafael Montes de Oca (naturalist), Ernesto Valdés y Melchor Calderón (assistants), Lauro Aguirre (secretary) y Joaquín Mendizábal Tamborrel (assistant calculator and draughtsman) (Urrutia 1964, pp. 138–139). This commission made proposals so that the governments of both countries could reach consensual agreement on the border.

Once the Exploratory Commission had completed their work, Mexico and Guatemala signed the Delimitation Treaty on 27 September 1882. Its first article states that Guatemala renounces any rights over Chiapas and Soconusco and the third specifies the frontier between the two countries, for which a scientific commission would erect markers on the ground. An important aspect of the delimitation treaty itself was that the agreement signed between Mexico and Guatemala in 1882 indicated the border would be defined by the rivers Suchiate, Chixoy or Salinas and Usumacinta. In addition to some key points such as the Tacaná volcano, Ixbul hill, two geodesic lines, a meridian (longitude 90° 59′29″ west) and three parallels (latitudes 16° 04′32″; 17° 05′08″ and 17° 49′, all north) and other key points whose coordinates the commissioners plotted. The Boundary Commission (*Comisión de Límites*) created by the 1882 Treaty was able to use the work of the previously mentioned Exploratory Commission.

On 1 October 1883, the Mexican Boundary Commission consisted of the following members: José Salazar Ilarregui (astronomer and leader of the commission), Próspero Goyzueta and Melchor Calderón (topographers 1st class), Reinaldo Gordillo and José H. Martínez (topographers 2nd class), Luis Servín (assistant, 1st class), Cristóbal J. Guillén (assistant, 2nd class), Rafael Montes de Oca (naturalist) and Fernando Flores (paymaster). Later, there were some later changes in the Mexican Commission members, and with the resignation of the engineer Salazar Ilarregui on 1 July 1884, Manuel E. Pastrana, who had joined as 2nd astronomer in October 1883, became the new leader (Pastrana 1914, pp. 166–167).

The Mexican Boundary Commission began its work on 1 December 1883, and because his health was poor, José Salazar Ilarregui insisted on Manuel Pastrana's appointment as the commission's 2nd astronomer; Salazar divided the Mexican Commission into five teams: the first, led by Salazar, drew the line from the river Suchiate to Ixbul hill. There, they determined a boundary marker called Vértice de Santiago, from which runs the first parallel that Goyzueta and Martínez drew by triangulation. It is defined by latitude 16° 04′ 32″. A fifth group, led by Gordillo, would go to finish mapping the rivers located between Tapizalá and Ixbul and, finally, Calderón would finish surveying the course of the rivers Coatán and Usumacinta. Salazar decided that Pastrana and Servín should go to Tenosique to astronomically determine some key points.

Several of the tasks performed by the Exploratory Commission were included in the final demarcation, as the Mexican geographical engineers had located the deepest channel of the rivers and astronomically fixed several points such as the summits of the Tacaná volcano and the Buenavista and Ixbul hills. For the definitive demarcation, the engineers in the field confirmed the deepest channel of the River Suchiate, mapping its course until they found the vertical plane that passes through the summit of the Tacaná volcano, and defined a geodesic line traced on the ground from the astronomically defined Vértice Muxbal marker to the Vértice Niquibil marker. From there, in accordance with the treaty, the frontier turns northeast as far as the Vértice de Santiago marker, from where the line follows parallel latitude 16° 04′ 32″ east until reaching the River Chixoy or Salinas. The line follows the course of this river until its confluence with the River Usumacinta where it continues along the deepest channel of the latter until reaching parallel latitude 17° 15′ 08″ north. The engineers drew a line on the ground that follows this parallel until it meets meridian 90° 59′ 29″, and at the intersection, they defined a point known as El Ceibo. From here, the borderline continues north until parallel 17° 49′, which was marked on the ground until reaching the River Azul, where the frontier with Belize begins. The groups erected boundary monuments, which also marked the end of their fieldwork.

With respect to the Mexico-Belize border, it is defined, mainly, by the deepest channel of the Azul and Hondo rivers. Part of the boundary is formed by an imaginary line in the sea, which divides the Bay of Chetumal and approximately follows parallel latitude 18° 10′ north as far as meridian longitude 88° 02′ west, continues along the meridian to the south and turns to the east to pass through the far north of Ambergris Caye.

Under Spanish colonial rule, Belize was part of the Yucatán peninsula. In the seventeenth century, pirates, for the most part British, French and Dutch, arrived at its coasts. The Spanish Crown granted them the right to settle on the land in an attempt to dissuade them from raiding Spanish vessels, but little by little, the pirates turned buccaneers took control of the islands, lagoons and barrier reefs. Attacks on ports and vessels continued and when piracy was outlawed, the pirates turned filibusters began to cut down hardwood trees to sell the timber. Now that they engaged in forestry, they imported slaves, and Spain could no longer expel them from the territory, which became recognised as a British colony under the name of British Honduras (De Vos 1993, pp. 67–72). As already mentioned, the Delimitation Treaty, known as the Mariscal-Spencer Treaty, was signed in 1893 and ratified in 1897 (AHSRE, 2005–7 f4). On 29 July 1897, an article was added to the Delimitation Treaty with Mexico that ensured freedom of navigation in British Honduran waters for Mexican merchant vessels. In 1898, Mexican Navy Commander Othon P. Blanco set out to found Payo Obispo (modern-day Chetumal) in order to establish a migration and customs post there that would control smuggling and achieve pacification in the area that had been afflicted for so long by the Caste War.

The Boundary Commission appointed for Guatemala continued working on demarcation of the frontier between Mexico and Belize, so they carried on measuring and marking parallel latitude 17° 49′ north on the ground, until they reached the meridian at Garbutt's Falls. This point determined the frontier between Guatemala and Belize. The borderline continues along that parallel until reaching the River Azul, whose course it follows to reach the River Hondo and its mouth near Payo Obispo, now Chetumal, in the Caribbean Sea. From here, an imaginary line continues through the sea, passing through the Bacalar canal to reach Ambergris Caye, known as San Pedro Island. Thus, the commissioners had to define the following points with geographical coordinates: (latitude 17° 49' north and longitude 89° 17′ 39.6″ west) the point where the River Azul meets the parallel of the dividing line with Guatemala; (latitude 18° 28′ 30″ north and longitude 88° 18′ west) the mouth of the River Hondo. They also defined parallels 18° 09′, 18° 10′ and 18° 25′ with imaginary lines in the sea. At these points, two lighthouses define the frontier: Rocky Point Lighthouse and Bulkhead Lighthouse (Fig. 5.3) (Tamayo 2015, pp. 127–133).

The Geographic Commission of the Republic of the Directory of Geographical and Climatological Studies drew the map based on the data provided by the Boundary Commission. The map, published in 1904, indicates the geographical position of Payo Obispo, the courses of the Azul and Hondo rivers and the maritime frontier.

The engineers confronted many problems on this southern frontier (Fig. 5.4), including climatic conditions, with rains and dense vegetation that hindered work, and other drawbacks that the members of Boundary Commission had to solve for themselves. Demarcation work was difficult and arduous, and both engineers and workers were prey to the illness and diseases typical of the region. Moreover, much of the information recorded in the field notes was lost in flash floods of the area's large rivers. Fieldwork ended in 1897, and the maps were completed in 1899.

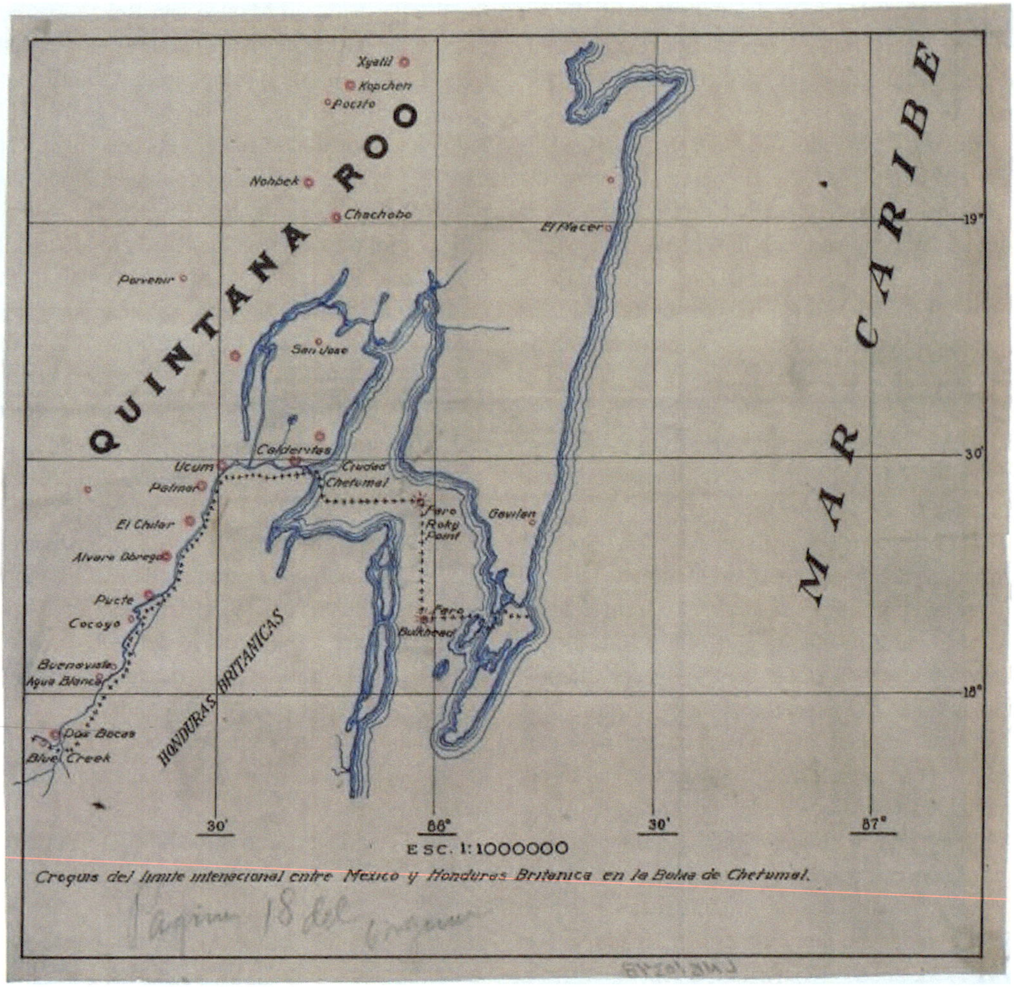

Fig. 5.3 Field plan of Chetumal Bay—frontiers between Mexico and Belize. *Source* MMOB, Límites entre México y Honduras británica/Boundaries between Mexico and British Honduras

5.4 Conclusions

Demarcation of a frontier is a significant responsibility as it indicates the boundaries of the territorial domain of a nation state. In the case of Mexico, this task fell to geographical engineers who, in the nineteenth century, possessed the knowledge of astronomical and topographic methods necessary to position permanent boundary monuments whose coordinates (latitude and longitude) it was possible to identify on a map and therefore receive acknowledgement as Mexico's international borders. The work conducted by the engineers who worked on these commissions was no easy task, although it was of the utmost significance for the country. The small group of Mexican scientists who worked on the two borders overcame all manner of obstacles, shortages, deficiencies, illness and disease, in addition to the deaths of fellow expedition members. Although their work has gone unacknowledged and unappreciated, it remains of great geographical value.

It might be assumed that relations between the commissioners were fraught, yet this was not the case, despite their apparent disparities. For instance, on the northern border, with differences

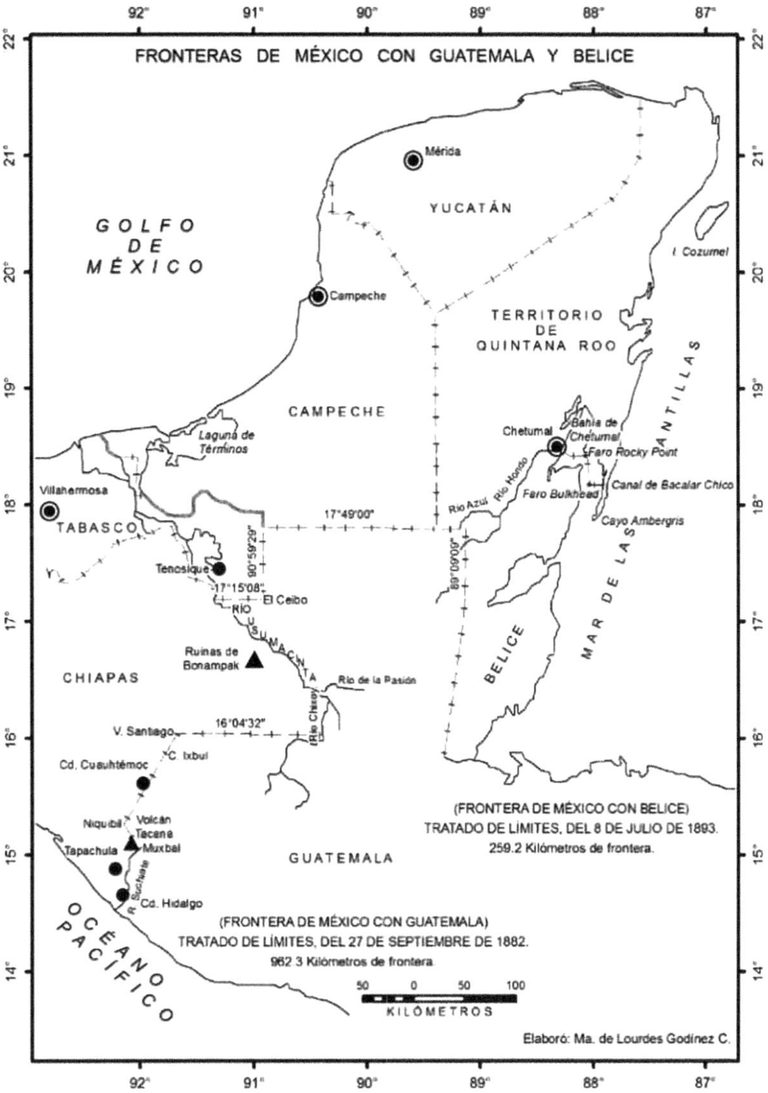

Fig. 5.4 Southern frontier—Border of Mexico with Guatemala and Belize. *Source* Tamayo 2014:151, based on MMOB and AHSRE

in scientific equipment, governmental support and escorts, at first the Mexican commissioners had to work almost to the point of exhaustion to prove their capability and keep up with their US counterparts. Although the US commission had more engineers, equipment and support, their scientific training was not superior to that of the Mexicans. The Mexican engineers were aware that they were protecting Mexican territory with their work, and that the marks on the ground would be the definitive ones from that moment on. On realising how well trained the Mexicans were, the US engineers offered to lend them their instruments, and over time their shared experience of hardship developed into a harmonious friendship.

Conversely, on the southern frontier, the residents of Chiapas had expressed their desire to remain part of Mexico, but Guatemala laid claim to the territories. The Exploratory Commission that had proposed the frontier later began demarcation of the border. However, work was hampered by the head of the Guatemalan Commission (Miles Rock, a US engineer, had a vested interest in the region and obstructed the progress of work). This continued until The Guatemalan government dismissed him and appointed Claudio Urrutia in his place. Urrutia was an intelligent, capable Guatemalan engineer who, alongside the other members of his commission, worked together with their Mexican counterparts to complete demarcation work. Demarcation of the frontier with Belize, at that time a colony of Great Britain called British Honduras, went ahead until an agreement was reached with the British government. The Mexican commission had continued surveying the rivers that form the borders between the two countries and finally the Exploratory Geographical Commission completed the cartographic work.

It is interesting to note that the scientific work of these commissions, in addition to scientifically establishing frontiers, meant that the country recognised the importance of cartography based on scientific methods. This made it possible for the engineer Agustín Díaz (Mexico-United States commission) to found and direct the Exploratory Geographical Commission (*Comisión Geográfico-Exploradora*), which was the first cartographic institution based on scientific methods. Under different names, this institution was responsible for the best maps in use until the middle of the twentieth century and was the forerunner of today's INEGI (Mexico's National Institute of Statistics, Geography and Informatics).

Vulnerability is no longer an issue—the demarcation of Mexico's borders is well established and fixed. However, the inequality between countries, together with the attraction that the United States of America holds for the poorer inhabitants of other countries (in particular those living in some regions of Mexico, Central American countries, Asia and Africa) means that thousands of migrants cross Mexican soil in endless caravans in the hope of reaching the powerful neighbour in the north. These migrants, motivated by the most basic needs, fleeing from persecution or removal from their lands of origin, hope to reach the United States. All too often, despite warnings, they fall foul of criminal groups, and unscrupulous delinquents recruit them. This makes the migrants extremely vulnerable because in many cases they are prevented from legally registering with authorities, who would protect them. The worldwide pandemic has only exacerbated this problem.

Archival Sources

AGN: Archivo General de la Nación. Mexico City.

AGN: *Tratado celebrado con los Estados Unidos de América, el 30 de diciembre de 1853 y ratificado el 31 de mayo de 1854*, Comisión de Límites México-Estados Unidos, expediente H-220 (72–73)/24, topográfica: 40-16-139.

AGN: Jiménez F (1857), *Diario Memoria de los trabajos científicos practicados bajo la dirección de Francisco Jiménez, 1er. Ingeniero de la Comisión de Límites Mexicana, conforme a las instrucciones del Sr. Comisionado don José Salazar Ilarregui*, Límites entre México y Estados Unidos, septiembre 1857. Inédito.

AHSRE: Archivo Histórico *Genaro Estrada* de la Secretaría de Relaciones Exteriores (Acervo Histórico-Diplomático). Mexico City.

AHSRE (1893): *Límites República Mexicana-Honduras Británicas, Informe del C. Ignacio Mariscal, secretario del Despacho de Relaciones Exteriores rendido ante el Senado acerca del Tratado de Límites entre Yucatán y Belice*. México: expediente 5002_7 caja 9, Secretaría de Agricultura y Fomento, Imprenta de F. Díaz de León, Sucursal Sociedad Anónima. inédito.

MMOB: Mapoteca Manuel Orozco y Berra. Mexico City.

MMOB Carpetas: *Límites México-Guatemala-Belice; Comisión de Límites México-Guatemala, Comisión Mixta de Límites*; Dirección General de Estudios Geográficos y Climatológicos (varias carpetas con documentos inéditos).

MMOB: Díaz A, Díaz L (1852) *Memoria sobre la topografía del río Bravo en la parte que comprende de su desembocadura a la villa de Laredo* (inédito).

MMOB: Diaz A, Díaz L (1853–54) *Memoria sobre la topografía del Río Bravo, en la parte que comprende de la colonia civil de San Ignacio a Los Amoles* (Inédito).

References

Anderson E (2003) International boundaries: a geopolitical Atlas. Routlege, New York

Berlandier L, Chovel R (1850) Diario de viage (sic) de la Comisión de Límites que puso el gobierno de la república bajo la dirección del Exmo. Sr. General de División D. Manuel Mier y Terán. Tipografía de Juan N. Navarro, México

Caballero F (2010) La Comisión Mexicana de Límites México-Guatemala, 1878-1899, Tesis para obtener el título de Licenciada en Geografía, Facultad de Filosofía y Letras. Universidad Nacional Autónoma de México, Ciudad de México

Ceballos RM (1989) La fundación de Nuevo Laredo (elementos para la interpretación de una tradición épica). Editorial Miguel Ángel Porrúa, México

De Vos J (1993) Las fronteras de la Frontera Sur. Reseñas de los proyectos de expansión que figuraron la frontera entre México y Centroamérica. Universidad Juárez Autónoma de Tabasco, Villahermosa (México)

Disturnell's Treaty Map (1848) Stagecoach Press, Santa Fe. Includes the Mapa de los Estados Unidos de Mejico, published by J. Disturnell, New York

Libura K, Morales L, Velasco J (2004) Ecos de la Guerra entre México y los Estados Unidos. Ediciones Tecolote, Conaculta (México)

Muratori, AM (2007) Fronteras: conceptos y caracterización espaciotemporal. In: López-Trigal L (ed) Fronteras Europeas y Latinoamericanas. De la geohistoria y los conflictos a los procesos de cooperación e integración. Universidad de León, León

Ortega y Medina JA (1989) Destino Manifiesto. Sus razones históricas y su raíz teológica. Editorial Patria, Alianza Editorial Mexicana y Consejo Nacional para la Cultura y las Artes, México

Pastrana ME (1914) Memoria sobre la cuestión de Límites entre México y Guatemala y sobre los trabajos ejecutados en la frontera de ambos países, por la Comisión Mexicana de Reconocimiento de dicha frontera primero y por la Comisión Mexicana de Límites para el trazo de la Línea Divisoria entre ambas Repúblicas, (cinco partes) México

Rebert P (2001) La Gran Línea. Mapping the United States-México Boundary, 1849–1857, University of Texas Press, Austin

Tamayo LMO (2005) La exploración de una frontera natural en el siglo XIX. Investigaciones Geográficas. Boletín Del Instituto De Geografía 56:134–159

Tamayo LMO (2015) La Comisión Mexicana de Límites y la definición de la frontera sur del país. Revista De Geografía Norte Grande 60:115–134

Tamayo LMO, Moncada O (2001) La comisión de límites de México y el levantamiento de la línea divisoria entre México y los Estados Unidos, 1849–1857. Investigaciones Geográficas. Boletín Del Instituto De Geografía 44:85–102

Tamayo LMO, Amendolla DC (2016) Idioma, comunicación y ciencia. Un caso de importancia nacional en el siglo XIX. In: Callicó J, González EJ (eds) Estudios y Testimonios en honor al Dr. Manuel Rodríguez Lapuente. Gobierno del Estado De Jalisco, México

Tamayo LMO, Garza G (2007)) Geohistoria de las fronteras latinoamericanas. El caso de México. In: López-Trigal L (ed) Fronteras Europeas y Latinoamericanas. De la geohistoria y los conflictos a los procesos de cooperación e integración. Universidad de León, León, pp 183–197

Tamayo LMO, Moncada O (2004) José Salazar Ilarregui 1823–1892. In: Armstrong P, Martin G (eds) Geographers, biobibliographical studies, vol 23, Continuum, New York–London, pp 116–125

Tamayo LMO (2001) La Geografía, arma científica para la defensa del territorio. UNAM, Instituto de Geografía & Plaza y Valdés editores, México

Tamayo LMO (2003) Una experiencia científica. Los ingenieros geógrafos en la frontera norte de México, 1849–1855. In: Berdoulay V, Mendoza H (eds) Unidad y diversidad del pensamiento geográfico en el mundo. Retos y perspectivas. Instituto de Geografía, UNAM - Instituto Nacional de Estadística, Geografía e Informática - Unión Geográfica Internacional (UGI), Comisión de Historia del Pensamiento Geográfico. México, Ciudad de México, pp 131–146

Tamayo LMO (2004) José Salazar Ilarregui, personaje central de la Comisión de Límites Mexicana, 1849–1857 y dos de sus colaboradores Francisco Jiménez y Agustín Díaz. In: Rodríguez ML (ed) Del estamento ocupacional a la comunidad científica: astrónomos-astrólogos e ingenieros (siglos XVII al XIX). UNAM, Ciudad de México

Tamayo LMO (2014) Las fronteras de México. Apuntes de su demarcación científica y técnica en el siglo XIX. Cuadernos de Geografía, Revista Colombiana de Geografía 23(2):139–157

Toussaint RM (1993) Belice, una historia olvidada. Instituto de Investigaciones Dr. José María Luís Mora, Centro de Estudios Mexicanos y Centroamericanos, México

Urrutia C (1964) Memoria sobre la cuestión de Límites entre Guatemala y México, presentada al Señor Ministro de Relaciones Exteriores por el jefe de la Comisión Guatemalteca, 1900. Centro Editorial José de Pineda Ibarra, Guatemala, C.A.

Vázquez JZ, Meyer L (1994) México frente a Estados Unidos (Un ensayo histórico, 1776–1993). Fondo de Cultura Económica, México

Zorrilla L (1977) Historia de las relaciones entre México y los Estados Unidos de América 1800–1958, tomo I. Editorial Porrúa, México

Part II

Contesting the Naturalness of Borders: Cartographic Practices, Geographical Knowledge and Territorial Issues

Borders on Paper, Borders on the Ground: Boundary Commissions Under the Treaty of Madrid and Their Role in the Reinvention of the Rio de la Plata Frontier (1752–1759)

6

Perla Zusman and Sergio Nunes-Pereira

Abstract

In this chapter, we will address the demarcation practices of the boundary commissions that, after the agreements reached under the Treaty of Madrid (1750), attempted to define on the ground the diplomatic decisions aimed at defining a distinct intercolonial border in the Rio de la Plata region. On the basis of the idea of associations by Bruno Latour, which assumes a principle of symmetry between materials and the heterogeneity of human and non-human properties, we explored the agency of hydrography and local societies in boundary demarcation. These agencies redefined not only the spatio-temporality of boundary commissions, but also the selection of their routes. Cartography, through inscription and translation operations, offers elements to identify these agencies and the different tiers of signification that challenged desk decisions. The controversy regarding the definition of the heads of the Ibicuy River proves, from our perspective, the creative nature of border reinvention.

P. Zusman (✉)
CONICET and University of Buenos Aires, Buenos Aires, Argentina

S. Nunes-Pereira
Fluminense Federal University, Niterói, Brazil

6.1 Introduction

This chapter will discuss the active role played by the boundary commissions that, after the signing of the Treaty of Madrid (1750), sought to delimit a border between the domains of the Portuguese and Spanish Crowns in the Río de la Plata region. This approach assumes that not only diplomatic agreements, but also the instructions received by the boundary parties were subject to resignification when the joint Spanish-Portuguese commissions met on the field.

As proposed by Gomes Castro (2012), we would like to highlight the agency of these boundary commissions. We will argue that the postulates of the treaties were put into practice, "transformed, altered, and translated and invented in multiple forms" (Gomes Castro 2012, p. 33) along the process of materializing desk decisions on overseas territories. Our work has been inspired by Gomes Castro insofar as he examined the tasks performed by the boundary commission in the north of Brazil by the mid-twentieth century. In spite of space and time differences, our study can take advantage of the heuristic elements in Gomes Castro's research project in terms of the bonds that the members of boundary commissions established with other (both human and non-human) actors during fieldwork and deskwork along the process of materialization of Brazilian borders.

J. García-Álvarez and P. Puente-Lozano (eds.), *Beneath the Lines*, Historical Geography and Geosciences, https://doi.org/10.1007/978-3-030-96904-2_6

Our interpretation is also based on the idea of associations[1] by Latour (2005), which assumes a principle of symmetry between materials and heterogeneous human and non-human properties. Associations between subjects, instruments, means of transport, the rules guiding demarcation, the characteristics of the environment, and the relations of convergence and conflict with local societies during fieldwork make the event of boundary demarcation a creative movement for border invention (Gomes Castro 2012, p. 30). This idea of association was also explored by Strandsberg when he considers that the ability to establish legal boundaries in the Artics depends "on a particular spatiality (...) it is space that decides how boundaries can be drawn and practiced." That's why he suggests "to investigate how space as a historical and social, yet material phenomenon is established" (Strandsbjerg 2012, p. 2).

This text will basically distinguish two types of agencies: the agency of river paths and that of local populations. These agencies will enter into associations with the actions of boundary demarcation expeditions, redefining the spatio-temporality of their activities along with their routes. From our perspective, cartography, in turn, by means of inscription and translation operations, will offer elements to identify these agencies and the different tiers of signification that challenged desk decisions.

This chapter will be divided into four parts: the first one will describe the purpose of the Treaty of Madrid for the Rio de la Plata region; the second will identify the division of survey areas between joint Spanish–Portuguese boundary commissions and the composition of these commissions. In the third part, we will examine the agencies of bodies of water and indigenous peoples to understand how they questioned the principles imposed by colonial authorities and the actions of commissions. Finally, we will analyze cartographic production as an element that will not only display the existing conflicts, but will have an active role in fabricating controversy and challenging the borders supposed to be demarcated as a result of negotiations.

6.2 The Purpose of the Treaty of Madrid in the Rio de la Plata Region

The interest in the definition of intercolonial boundaries between the territories of Portugal and Spain in South America may be inscribed against the background of eighteenth-century international relations. During this period, the Iberian Crowns sought to improve the management of their resources and the control of the geographical environments under their dominance. With this in mind, they performed changes in their political administration and fortification, militarization and colonization actions on overseas territories.

In line with the above, the Treaty of Madrid between the Spanish and Portuguese Crowns features special traits. The only agreement that until then distributed territorial possessions between Portugal and Spain was the Treaty of Tordesillas, dating from 1494. In spite of it being constantly invoked in negotiations, this treaty had become obsolete for several reasons. First, the Pope's role to mediate in territorial political conflicts had lost relevance. Second, there was more information available on overseas possessions in the eighteenth than in the fifteenth century. Third, the search for markets in the Rio de la Plata area by other European powers aroused a stronger interest in Spanish and Portuguese possessions than in the 1500's. Irrespective of the conflicting relations that could exist between both colonial powers, defining their areas of dominance and control protected both Crowns against the ambitions of other European empires

[1]Latour (2005) points out that the actor-network theory relies on a sociology of associations, i.e., it highlights the relationship that is established between heterogeneous elements. "These elements include (...) human beings, the significations we produce, symbols, discourses, but also material elements, objects, technical artefacts and items" (Tirado Serrano and Domenech i Argemí 2005, p. 10). Conversely, a sociology of social issues, which is the major concern of social scientists, will prioritize the study of bonds between objects and homogeneous situations.

(Zusman 2000; Herzog 2015). At the same time, Spain and Portugal aimed at establishing a form of territorial organization that would prevail over pre-existing forms like Jesuit Missions or those typical of indigenous peoples, in addition to tightening control over smuggling through Colonia de Sacramento (Moutoukias 1988; Cavalcante 2006; Possamai 2014).

The Treaty of Madrid reflected the criteria agreed by consensus during negotiations: on the one hand, *utis possidetis* and, on the other, the natural frontiers, i.e., the definition of intercolonial boundaries based on a selected natural element (the source or the path of a river or the hills considered to be "most prominent"). It was assumed that a natural frontier criterion would ensure a clear and precise definition of boundaries and that these would not be altered over time (Sahlins 1987).

In addition, the *Mapa de los confines de Brasil en las tierras de la Corona de España en América Meridional* [Map of the confines of Brazil with the lands of the Spanish Crown in South America], known as *Mapa de las Cortes* [Map of the Two Courts] (Fig. 6.1), played a critical role. Considered to be a complement to the treaty, a graphic materialization of the results of the negotiation, it would be used as a guideline to fieldwork by the Spanish and Portuguese joint boundary commissions.

In the Rio de la Plata area, the Treaty of Madrid involved mainly the Portuguese ceding of Colonia de Sacramento and, in return, the Spanish transfer to Portuguese control of the territory to the east of the Uruguay River. The latter would entail the transfer of the seven Guarani Missions located in the area.

6.3 Creation of the Spanish–Portuguese Joint Commissions

Under the Treaty of Madrid, boundary demarcation in the South American territory would be conducted by a northern and a southern unit. While the former would work in the Amazonian area, the latter would fulfill its activity in the Rio de la Plata region. Each division, in turn, would split into three parties that would materialize on the ground the imaginary line agreed by the Crowns under the treaty (Martínez Martín 2000; Martín-Merás 2007). The activity of the three parties would be commanded and supervised by High Commissioner Gomes Freire de Andrade on behalf of Portugal and by the Marquis of Valdelirios on behalf of Spain (Figs. 6.1 and 6.2).

Pursuant to the provisions of article IV of the treaty regarding the Rio de la Plata region, the first party should demarcate the boundary from Castillos Grandes up to the Ibicuy River. The commissioners of this party would be Francisco Antonio de Meneses e Sousa on behalf of Portugal and Navy Captain Juan Echavarría on behalf of Spain. Work started in 1752, but was interrupted the following year as a consequence of the breakout of the so-called Guarani War.[2] Fieldwork activities were not resumed until 1758. This time, the Portuguese demarcation was commanded by José Custódio de Sá e Faria. The main problem now was associated with the difficulty to identify the heads of the Ibicuy River: were these horizontally aligned to the stream that empties into the Uruguay River, or, on the contrary, did the Ibicuy heads originate at the southeast and run diagonally until they flowed into the Uruguay River? We will analyze this problem in more detail in Sects. 6.4 and 6.5.

The second party began to work in 1759. According to article V of the treaty, demarcation would be conducted from the Ibicuy River up to the Igurei river course on top of Parana River's Salto Grande. Then, the dry frontier would be defined up to the sources of the San Antonio River. The commissioners of this party were José Fernandes Pinto Alpoim representing Portugal and Francisco de Arguedas on behalf of Spain. This time, the party's activities were halted as a consequence of indigenous rebellions. In spite of the difficulties posed by the definition of the

[2]The Guarani War is known as a series of military confrontations that took place from 1754 to 1756 where indigenous troops, supported by Jesuit fathers, resisted the advance of Portuguese and Spanish demarcation parties and were defeated by a Spanish–Portuguese army at the battle of Caibate (Quarleri 2009; Tau Golin 2014; Erbig 2015).

Fig. 6.1 Map of the Two Courts, identifying each area where the boundary commissions of the Southern Region should work. *Source* Authors based on the Mapa de los confines de Brasil en las tierras de la Corona de España en América Meridional [Map of the confines of Brazil with the lands of the Spanish Crown in South America], 1749. BNE

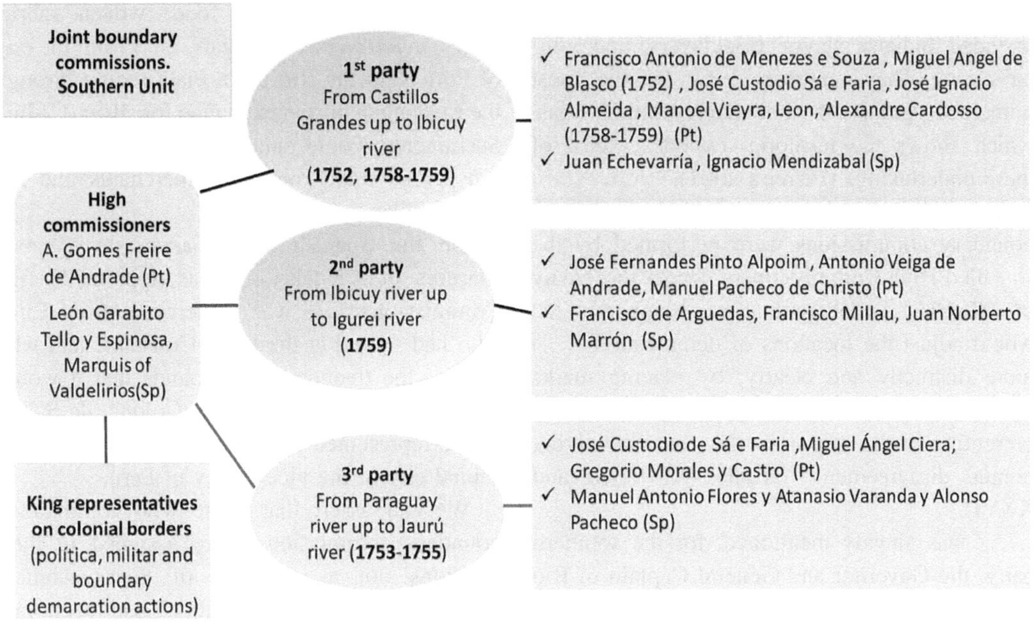

Fig. 6.2 Chart identifying the commissioners and experts of the three boundary commissions in the Southern Region. *Source* Authors' own elaboration

Santo Antonio and Pepirí river courses, the controversy was solved and demarcation tasks ended with the production of "a general map"[3] (Ferreira 2001, p. 172).

The third party would work from the Paraguay to the Jauru rivers. The commissioners in charge were José Custódio de Sá e Faria on behalf of Portugal and Manuel Antonio Flores representing Spain. This party, which was less stricken by indigenous conflicts, attained the demarcation goals between 1753 and 1754. Certain difficulties arose, however, to delimit the final section, as "it was not possible to go all the way downstream the Aguarahy river on account of waterfalls, extremely dense vegetation and the presence of Monteses Indians" (*Diário de la terceira partida de demarcação de América Meridional* [Diary of the Third Boundary Party in South America] in Ferreira 2001, p. 185). High Commissioners Gomes Freire and Valdelirios considered that knowing the location of the head of the Aguarahy River (currently Ipané) and the point where it entered the Paraguay River was enough to complete demarcation, and that it was unnecessary to know the full path of the river course (Ferreira 2001).

Each party was formed by three different kinds of members: first, those who performed political tasks; second, those who focused on technical jobs; finally, those whose activities included protecting the members of the boundary commissions, carrying packages, acting as vessel rowers, opening up paths in the wilderness, taking care of the food and clothes of commissioners and technical experts. While the first group represented the royal authority on these distant territories, the second group was formed by officers, military engineers, cartographers, astronomers, and designers, in addition to

[3]This is the *Mappa da demarcação que por orden de S.M. F e C fez no anno de 1759, a Segunda Partida da boca do Rio Ybicuy, seguindo o curso dos Rios Vuruguay, Pipiry, S. Anto, Yguacú e Paraná thé o seu Salto, Povo de S. Nicolláo, a 8 de Abril de 1760* [Map of the demarcation that by order of His Most Faithful Majesty and His Catholic Majesty was drawn in 1759 by the Second Ibicuy River Mouth Party following the path of the Uruguay, Pipiry, San Antonio, Iguazu and Parana rivers until Salto Grande]. San Nicolás village, April 8, 1760, produced by Manuel Pacheco de Cristo. Available from Itamaraty Map Collection, Rio de Janeiro, no 370., fig. XI.

surgeons and priests. Finally, the third group included Indians, slaves, bricklayers, and military men. They represented by far the most numerous group in boundary commissions, which shows the territorial conquest spirit of these undertakings (Ferreira 2001).

The political tasks to be carried out by the boundary commissions were performed by the so-called High Commissioners. Under the Treaty of Madrid, "intelligent commissioners (...) would adjust the locations of demarcations (...) more distinctly and clearly, by placing marks where they deemed it most appropriate (...) and presenting to the monarchs the issues involving mutual disagreement" (articles XI, XXII and XXVI).

As was already mentioned, for the southern party, the Governor and General Captain of Rio de Janeiro Gomes Freire de Andrade, future Count of Bobadela, was appointed High Commissioner by Portugal. Having been in office since 1733, he was the most knowledgeable person about the south-east and southern regions in Brazil.[4] He disagreed with the resolutions under the Treaty of Madrid and had more affinity with the views of the future Marquis of Pombal, for whom Colonia de Sacramento had more strategic and economic value than the lands occupied by the seven Missions. Moreover, Gomes Freire advocated the positions of Rio de Janeiro merchants, who imported multiple goods (manufactured products, food, woven fabric, wines, *aguardientes*, vinegars, etc.) from the city of Porto into the Rio de la Plata region through the Lusitanian port in exchange for silver. Ceding Sacramento would entail closing a thriving and prosperous trade route, for merchants and the people alike.

On the Spanish side, Gaspar de Munive, Marquis of Valdelirios,[5] was appointed High Commissioner. He was a merchant from Lima who had served at the Lima Consulate and who favored the treaty on the grounds that it would end with the competition that Colonia de Sacramento represented for mercantile activities in the capital city of the viceroyalty of Peru.

We can assert that those who commanded boundary commissions were assigned to their positions not as a result of their technical expertise, but of their political and economic acquaintances. Their networks and the activities these royal officers conducted were precisely what enabled them to perform these roles. In brief, the boundary commissioner was also a sort of Crown representative, both on the border and when implementing any actions that could impact border definition.

They had to impose the King's authority over any other form of exercise of power in the area, whether originating from the missionaries or the viceroys. It could then be said that, in the absence of the King, the head of a boundary commission carried out the actions that, in medieval times, monarchs performed along the European borders (Sahlins 1993, p. 46). Now, supported by scientific criteria, the commissioner of a boundary expedition would solve controversies and forge alliances. In addition, in the Rio de la Plata area,

[4]Gomes Freire pursued a military career and became a major officer of the Portuguese cavalry. He was commissioned to Brazil to hold key positions, first at the General Captaincy of Rio de Janeiro (1733) and then in Minas Gerais (1735), during the mining boom. In 1738, the Captaincies of Santa Catarina and Rio Grande became subordinate to that of Rio de Janeiro; therefore, the Southern border became under its control, in practice. He was a strong advocate of Portuguese territorial expansion policies over the Rio de la Plata region. This led him to participate directly in the defense of Sacramento and planned the establishment of the Rio Grande de San Pedro enclave in 1737, when Colonia was besieged by the Spaniards. With similar objectives in mind, he advanced the opening up of roads between the South and São Paulo, the establishment of cities and villages in Rio Grande, the development of fortifications in Rio Grande and Santa Catarina Island, and canals to control the Los Patos and Mirim lakes.

[5]Gaspar de Munive León Garabito Tello y Espinosa, Marquis of Valdelirios, was born in Huamanga (Peru viceroyalty). He studied at the Real de San Martín Jesuit school. He received the *corregimientos* Huanta and Cañete at the viceroyalty of Peru from the court of Philip V. He was recognized as a wealthy merchant from Lima. During the *Asiento* war against Great Britain (1739), he was made prisoner by the British. Upon being released, he was incorporated to the Council of the Indies in 1748. In 1751, he was appointed Spanish plenipotentiary commissioner to perform demarcation tasks under the Treaty of Madrid.

they took the initiative to prepare for war when the continuity of demarcation actions was blocked by the advance of indigenous peoples.

6.4 The Actions of the First Party and the Conflict Over the Definition of the Ibicuy River Border: The Agency of Rivers and Local Populations

As we mentioned in previous paragraphs, the actions conducted by the first party would define the boundary over an area comprised from Castillos Grandes up to the Ibicuy River. In February, 1753, the actions conducted by the members of the boundary commission were stopped by the presence of an indigenous militia close to Santa Tecla. Therefore, they left the region and demarcation had to be conducted in two stages: 1752–1753 and 1758–1759. While in the first stage the mixed Spanish–Portuguese commission worked on the erection of boundary stones from Castillos Grandes Bay up to de los Reyes Hill,[6] during the second stage this commission covered from Santa Tecla up to the Ibicuy river mouth into the Uruguay River.

According to Ferreira (2001, p. 152), the definition of the source of the Ibicuy River was the largest difficulty encountered by the parties to agree on an intercolonial border. From our perspective, the agency of rivers and local populations establish associations with the means of transport (smacks, canoes) and with the instruments of observation, orientation, measurement, and drawing (theodolite, sundial, compass, quadrants, telescope, microscope, barometer, thermometer, notebooks, and brushes) (Gesteira 2014). These associations were resignified to construct arguments based on scientific support in favor of the Spanish or Portuguese positions.

In fact, as per the Spanish commission, even if the border was cartographically represented on the Map of the Two Courts received by the parties, the treaty was not explicit about the criterion that should be used to join the heads of the Negro and Ibicuy rivers.

> Article 4 of the Boundary Treaty (…) states that beyond the Negro river heads [the divide] will continue up to the main source of the Ibicuy river: going downstream up to its mouth into the eastern bank of the Uruguay river: with no explanation as to what means we should use to reach other sources, of any other river (…).” (Don Juan Echavarria to Francisco Antonio Cardoso de Meneses y Souza, Santa Catalina, May 18, 1758. AGS, Estado, 7402)

This information was complemented by data found on the Map of the Two Courts showing that the boundary between the Negro and Ibicuy rivers would be defined by prominent rising landforms dividing the waters toward the Uruguay River and the Atlantic Ocean (Fig. 6.3).

The sources of the Ibicuy River then began to be discussed in this context. For the Portuguese party, the heads of this body of water would be located in Santa Tecla (which would be consistent with the demarcation established on the Map of the Two Courts). On the contrary, the Spanish party claimed that these sources could be located farther north, at Monte Grande. The Spanish proposal would result in a review of the imaginary line advocated by the central governments (Fig. 6.6).

The possibility that the Ibicuy river heads were located at Monte Grande Hill endorsed one of the criteria that ensured clarity in the definition of the boundary: the body of water emptying into the Ibicuy River should form a straight line. Precisely, the river considered to be the Ibicuy Mini, originating at this hill, could be used to support the divide.

According to Spanish cosmographers and cartographers, the Iberian Crowns did not know about the existence of these rivers before the armies crossed the areas. This information was privileged for the Jesuit priests and the indigenous peoples, and it needed to be incorporated to the cartography produced by the parties. The Portuguese, on their part, favored the idea of marking on the ground the delimitation defined

[6] Boundary stones were placed in Castillos Grandes, India Morta y Serro dos Reis (Ferreira 2001).

by the instructions and the map issued by the courts, regardless of the name of the river supporting the line.

The Portuguese argument position was clearly explained in a set of manuscripts collected by Miguel Angelo de Blasco in order to rebut the Spanish position.[7] These manuscripts include expert opinions, maps, and letters exchanged between Colonel de Blasco; the Spanish plenipotentiary commissioner, the Marquis of Valdelirios; and his geographers Francisco Millau and Francisco de Arguedas. From the examination of these documents, it can be inferred that de Blasco wished to demonstrate that the body of water identified at Monte Grande (Ibicuy Mini), which the Spaniards proposed to use to "correct" the boundary, could not be found on any ancient or modern map and that, therefore, it did not prove the existence of any other Ibicuy River different from "the one instructed to be demarcated by the sovereign contractors" (*Juízo ou parecer do Coronel de Blasco sobre uma carta de 9 de abril e pequeno mapa remetido do Sr. Marquês de Valdelirios.* [Colonel de Blasco's judgment or expert opinion about his letter dated April 9, and small map forwarded by the Marquis of Valdelirios]).

In de Blasco's terms, ancient and modern maps meant those compiled by Jesuit cartographers, those resulting from the activities of the first party, and the "geographical maps printed far and wide" that had been used as a basis for the Map of the Two Courts. The question is analyzed in more depth in the *Post scriptum do Coronel de Blasco, em 16 de junho, em resposta ao Geógrafo D. Francisco Millau Maraval* [Postcript by Colonel de Blasco, June 16, in response to geographer D. Francisco Millau Maraval] attached to the same set of documents.

In his writings, de Blasco wondered whether the indigenous narratives, which, in his view, were contradictory, were enough "to refute what scholar priests and brave geographers had done from the beginning up to the present guided by pilot Indians" (*Juízo ou parecer do Coronel de Blasco* ... [Colonel de Blasco's judgment or expert opinion…]). In regard to the Spanish map that introduced Ibicuy Mini into the scenario, Colonel de Blasco stated that it disregarded all the geography charts produced until then, barely based on:

> a couple of simple Indians and a priest (…), who [would receive] more credit than all the scholars who worked in the Geography of this America, on whose ancient and modern maps the Ibicuy river is shown in conformity with the maps forwarded by both contractor Courts. (Juízo ou parecer do Coronel de Blasco [Colonel de Blasco's judgment or expert opinion…])

On these grounds, in de Blasco's view, the legitimacy of the Spanish proposal was feeble and he even deemed it impertinent—"the Sovereigns did not mandate that the river be deprived of its name (…), least of all that the name be changed" (*Juizo o parecer do Coronel de Blasco…*[Colonel de Blasco's judgment or expert opinion…])—so the members of the boundary commission were to follow the royal instructions to the letter to complete the demarcation line as fast and as accurately as possible.

The controversy was intended to be solved by discussing the three questions below: the names of the rivers on Jesuit cartography, the definition of the mainstream of the Ibicuy River and the occupation of the disputed territories by the indigenous peoples. The commissioners of both parties relied on the Jesuit cartography produced in earlier times to identify the Ibicuy's river path as drawn on the Map of the Two Courts and what river path the indigenous peoples called Ibicuy.

[7]Blasco, Miguel Angelo de. *Notta dos papeis, q. vão no masso junto, feitos pelo Coronel de Blasco, por ordem do Exmo. Ml. de Campo Gen.al e Comissr^{o} Plenipotenciario Gomes Fr.e de Andrada, em resposta às questoens novamte. sucitadas do S. Marquez de Valde Lirios, e seus geographos, a respeito da demarcação de Primr^{a} Partida, no acto de continuala neste anno de 1758.* [Note to documents going in mass together written by Colonel de Blasco by order of His Excellency Field Marshal General and Plenipotentiary Commissioner Gomes Freire de Andrada in response to questions raised again by the Marquis of Valdelirios and his geographers regarding the demarcation by the First Party upon the continuation thereof this year, 1758.] 58 folios, 4 color maps, 49 cm. Biblioteca Nacional do Brasil. Manuscritos e Cartografia. Document 24. I-28, 24, 014. Available at: http://objdigital.bn.br/acervo_digital/div_cartografia/cart1095063/cart1095063.pdf.

In this manner, the question about the location of the river sources could be solved. In fact, however, certainties and consensus would only apply to the river's lower course, near the mouth into the Uruguay River. Unfamiliarity with the terrain and the complexity of the hydrographic network sustained the construction of different hypotheses consistent with the territorial ambitions of each Crown. In spite of that, not all available cartography named the same river course Ibicuy; in addition, complementary denominations suggested that the source could originate in minor streams across different areas.

For example, while on father José Quiroga's Map used by the Spanish commissioner

> two branches were located at its heads, one unnamed coming from the South, the other called Ibicuy Mini running east-west (…) the Portuguese commissionaire looked at the map drawn by Tadeo Enis, where a southern branch called Ibicuy Guazu came from Santa Tecla and collected the smaller streams that formed it at the source like Ibicuy Rapi close to Santa Tecla, and Tacuarembo, coming downstream from Monte Grande. (Martínez Martín 2000, pp. 439–440)

The other criterion that was taken into account was the possibility of defining the main course of the Ibicuy River after exploring its flow. Climate conditions and a lack of consensus as to the form of conducting this task made it impossible to solve the controversy. First of all, it was impractical to measure the flow of the water course recognized on the Map of the Two Courts because, when some party members crossed the "ancient" Ibicuy, they saw that "the rainfall that was abundant in December and January having failed, in March the river called Ibicuy Guazu had less water (…) it was almost dry when it was crossed" (Marquis of Valdelirios to Ricardo Wall, Pueblo de San Nicolás, May 16, 1758. AGS, Estado, 7402).

Second, the Portuguese party did not agree with Echavarria as to exploring the Ibicuy source. According to Miguel Angelo de Blasco, Echavarria conducted a "bottom-top" demarcation (from the Ibicuy's mouth toward the river head), while a more suitable method would have been to advance in the opposite direction, "from top to bottom" (from the river heads to the mouth), following the river path.

Valdelirios, in turn, claimed that the way how the river courses were explored had been agreed upon with the Portuguese High Commissioner Gomes Freire.

> As D. Gomes Freire changed the assumption on which I was based (…) hence he agreed with me (…) that the examination of the source and main head of the Ibicuy river would be started at the river mouth, when D. Juan de Echavarria arrived, and, to facilitate the enterprise, he judged that, as there were no doubts about the main body of this river, it would suffice to examine, from the confluence or junction formed by the two main river arms, which one brought more water to describe by this trait the head and main source of the Ibicuy river, which is ordered to be found in chapter 1 of the Treaty. (Valdelirios to Wall, Campo de Santa Catalina, October 20, 1758. AGS, Estado, 7402)

In the referred to documentation de Blasco included a map at a scale of 20 leagues drawn by the Spanish geographers and forwarded to the Portuguese that summarized in graphic terms the incompatibility between the Spanish and Portuguese perspectives (Fig. 6.4).

For the purposes of this analysis, our main interest lies not so much in the principal map including the demarcation lines advocated by each commission, but the other, more detailed maps to the left and to the right, designed by the Portuguese geographers. These maps include "tokens" of the "correct" manner to conduct the demarcation from top to bottom along the river course, "even when instructions mandate otherwise" (Example V). However, exceptions are accepted, as in the case of tributaries that empty into the Ibicuy river, which would be "demarcated upstream and not downstream" (Example G). Anyway, the examples highlight the problem of "changing the manner of demarcation," as suggested by the Spaniards. This is considered to be "improper and uncomfortable and against Royal determination," according to the legend in Example V. Along these lines, de Blasco seemed to have no doubts about the point where flow measurements should be started. Thus, in the bottom portion of the major map, to the left of the scale, the following can be read: "And—the

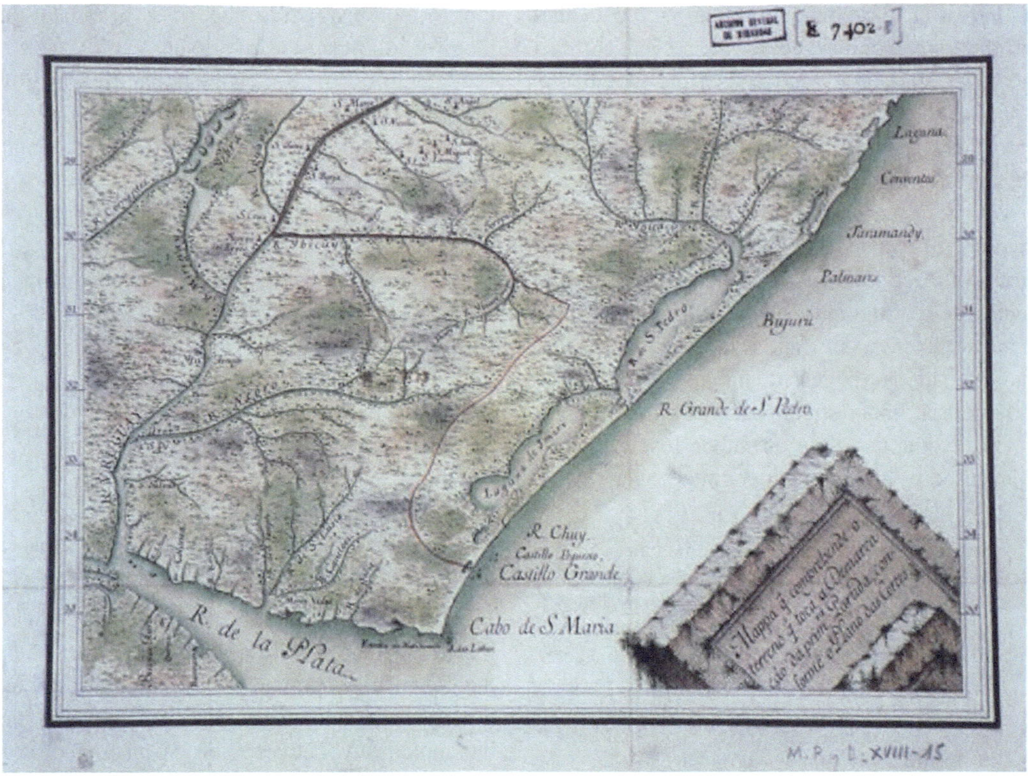

Fig. 6.3 *Mappa que comprehende o terreno que toca a demarcação da Primera Partida conforme a Plano das Cortes* [Map of the land to be demarcated by the First Party according to the Map of the Two Courts]. *Source* AGS. Mapas, Planos y Dibujos, 06, 003

summit that empties its waters into the Negro river, into the Ibicuy river, on top of which the main head of the (…) river should be found as stated in Art. 1 of the Treaty." In this way, the demarcation line would follow the definition established in the documentation prepared by the central governments.

Beyond this methodological criticism, Miguel Angelo de Blasco wonders how the route of the demarcation line would be defined from Santa Tecla to the heads in the Missions in case the Ibicuy river heads were located at Monte Grande Hill.

The occupation of the area under conflict is the last criterion we would like to underscore, introduced by the Spanish Crown, although covertly. The area in the vicinity of Santa Tecla had not been evacuated after the Guarani conflict, inasmuch as it was considered that the demarcation line was to be oriented eastward instead of southward. The Indians of the ancient mission of San Miguel had moved to a location close to the Ibicuy River where they had their *estancias* (estates) that formed San Nicolás village.

> the incertitudes that arose upon searching for the main source of the Ibicuy river leave the question of whether these Indians should move or not unresolved, because the land they occupy may belong to the Portuguese Crown, the demarcation of which H. M. believed to have been completed by the first party commanded by D. Juan de Echavarria, at the time when H. M. took care of collecting the scattered Indians or at least finding the main source of the Ibicuy river, for which a few days were left, and, by deferring it for such a long time that the continuation of the demarcation interrupted in the year 1753 in the proximities of Santa Tecla has not yet been restarted, H.M. is now faced with the pressing need of choosing a disposition toward one of two extremes (…), the one involving the suspension of the Indians' move

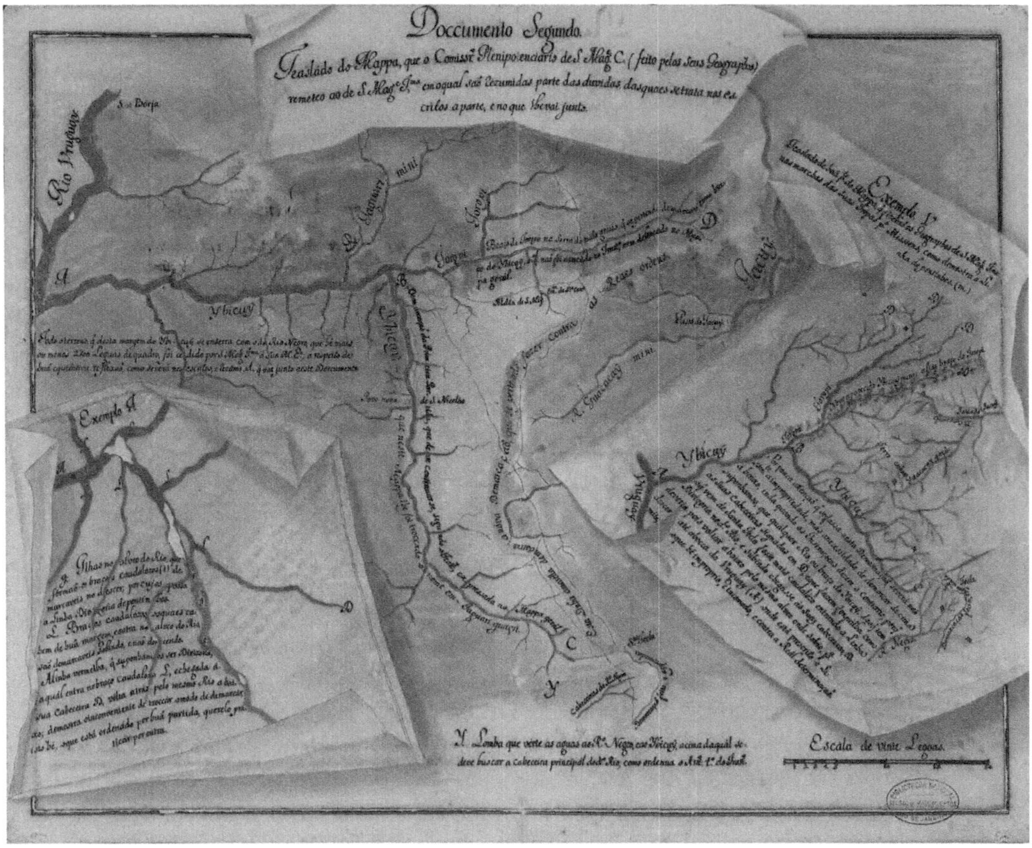

Fig. 6.4 *Traslado do Mapa que o Comisario Plenipotenciario de Sua Majestade Católica (feita pelos seus geógrafos) remiteu a sua Majestad Fidelísima, em o qual sao resumidas parte das dúvidas, das quaes se trata nos escritos aparte, e no que elevai junto*[Reproduction of the Map sent to His Most Faithful Majesty by the High Commissioner of His Catholic Majesty (drawn by his geographers), where a portion of the incertitudes is summarized, which are dealt with in the separate writings, hereto attached]. *Source* Document 24. I-28, 24, 014. BNB. Manuscritos e Cartografia

> until a resolution is reached regarding to which of both Crowns the occupied land belongs, in which case, if belonging to Portugal, H.M. claims they will be ordered to leave (…) and the other one being (…) the case that H.M. should command they move forthwith. (Cevallos to Valdelirios, Paso del Yacuy, August 3, 1758. AGS, Estado 7402)

The Buenos Aires governor, Pedro de Cevallos, in disagreement with the evacuation of the lands occupied by the Missions, recommended maintaining the settlements on the left bank of the Ibicuy River to avoid "oppressing the peoples by the Parana and the western margin of the Uruguay rivers" (Cevallos to Valdelirios, Paso del Yacuy, August 3, 1758. AGS, Estado 7402). In his opinion, the forceful migration of these peoples could give rise to a new rebellion, especially because in the areas where they should be forced to settle there would be no livestock or agricultural products for food. In addition, the occupation could be used as a criterion for these lands to remain under Spanish dominance, as was actually the case.

The Lusitanian Crown, in turn, using the secret arguments recommended by the future Marquis of Pombal, attributed the delay in the evacuation of the seven Missionary peoples to the unwillingness of the Spanish Crown to comply with the Treaty, since, out of the 30,702 inhabitants who lived there, just slightly more than 14,000 had been moved to the western margin of the Uruguay river, with 16,000 still remaining to be moved to the *estancias* (Valdelirios to Cevallos, Santa Catalina, October 6, 1758. AGS, Estado 7402). The Portuguese recognized that the lands located south of the Ibicuy River were occupied by "70 families composed of the peoples of S. Angel, S. Lorenzo, which were later joined by other families from San Nicolás and San Juan" (Valdelirios to Ricardo Wall, Pueblo de San Nicolás, May 16, 1758. AGS, Estado, 7402). In the meantime, Valdelirios was also mistrustful of Lusitanian authorities, as he perceived little interest from their part to comply with the treaty.

The difficulty to solve the conflict about the Ibicuy river heads made the parties leave the matter in the hands of the commissioners, who, in turn, referred the issue to the sovereigns of both Crowns. The reasons to abandon the efforts to solve the controversy might be understood by the fact that this was one of the most strategic areas, linking Colonia de Sacramento to the entry to the Rio de la Plata territories. From a more local perspective, Portugal's ambition was to expand its domains inland and occupy the lowlands of the current Rio Grande do Sul, considered to be fertile and very good for livestock. In turn, the Spaniards wished to keep the lands occupied by the migrant Indians and did not intend to give control to the Portuguese of any more territories (Ferreira 2001, p. 205).

6.5 Cartography: Inscription, Translation, and Expression of the Heterogeneous Association that Would Reinvent the Border

The controversy about the Ibicuy river source would lead to the production of abundant correspondence and cartography by the officials participating in field activities, between them and high commissioners, and between these King representatives on the border and central government authorities. It was a territorial problem which, expressed into a technical language and supported by scientific criteria, sought to avoid evidencing any signs of fragility. Therefore, the protagonists were professionals who, with refined technical expertise, such as de Blasco or Custódio de Sá e Faria, on the Portuguese side, and Arguedas and Millau y Maraval, on the Spanish side, could actually develop the necessary rationale to legitimate the position advocated by each Crown, mainly expressed by the letter exchange between high commissioners Valdelirios and Gomes Freire de Andrade.

In this section, we will draw particular attention to the cartographic efforts based on fieldwork conducted by the abovementioned experts. We can distinguish different types of maps: from those on which we identify an initial map (like the recreation of fragments of the Map of the Two Courts) to those that are an overall result of the completed tasks (maps at a small scale), including other fragmentary maps (at a larger scale) that are the product of surveying the explored areas. Many of these cartographic products combine "rational" with "esthetic" representations, and may have the purpose of either being used in situ by boundary parties for ground demarcation or proposing a work route oriented to clarifying or solving the controversy. In the latter case, maps were handed over both to high commissioners and to the ministers of the courts.

We are particularly interested in highlighting the significance of three maps (Figs. 6.4, 6.5 and 6.6).

These three maps may be considered to be inscriptions (Latour and Woolgar 1979) of the conflict. On the one hand, they were developed on the basis of knowledge acquired in the field with the help of all the instruments used in mapping processes. In this connection, it is interesting to highlight the dialogue with other maps (like the Map of the Two Courts or those produced by the Jesuits) and with knowledge from indigenous peoples. In fact, many streams regained the name given to them by local populations. Thus, reference E in the explanation of the map developed by Custódio de Sá e Faria *Demostração do Rio Ibicuy e os bracos que o formão* [Demonstration of the Ibicuy river and the river arms that form it] (Fig. 6.5) recognizes

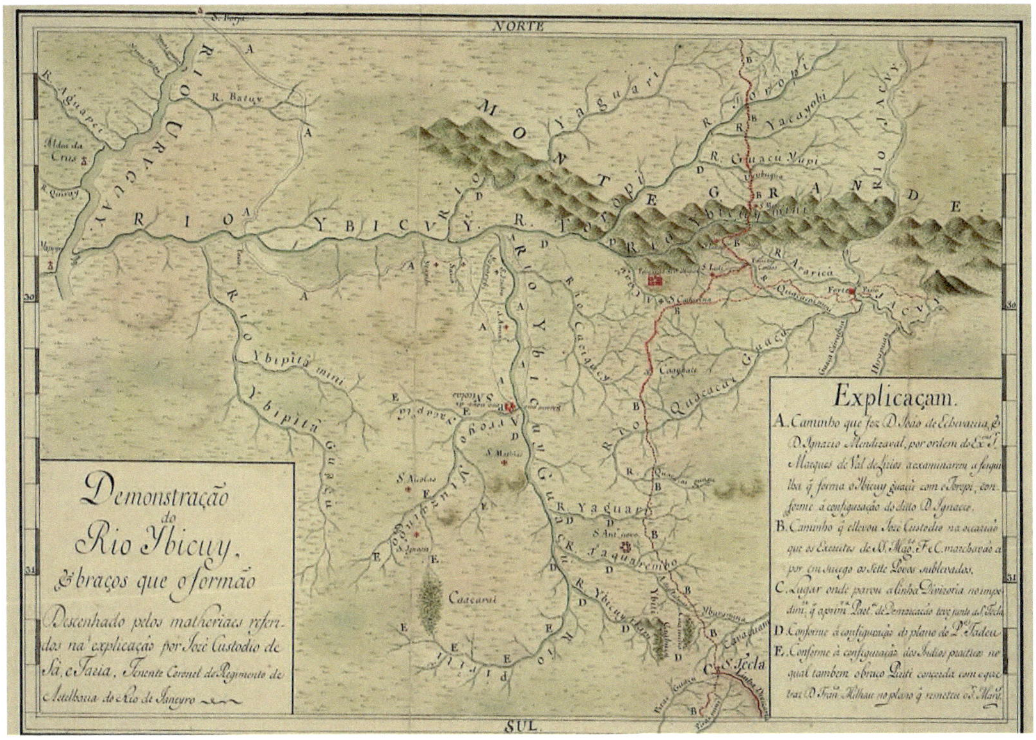

Fig. 6.5 *Demostração do Rio Ibicuy e os braços que o formão. Desenhados pelos matheriaes referidos na explicacão por José Custódio de Sá e Faria, Tenhiente Coronel de Regimento de Rio de Janeyro* [Demonstration of the Ibicuy river and the river arms that form it. Designed by the materials referred to in the explanation by José Custódio de Sá e Faria, Lieutenant Colonel of the Rio de Janeiro Regiment]. *Source* AGS. Mapas, Planos y Dibujos, 06, 002

that the name "Ybicuy Mini" was given by "the Indians from Pueblo Nuevo de San Nicolás (…) to the river at the foot of the settlement (…) by the Jaguary river."

On the other hand, however, these three maps exchange a dialogue between them and actively fabricate controversy. While the first one (Fig. 6.4) places the conflict against the background of the delimitation process that the first demarcation party was supposed to complete, the second (Fig. 6.5) presents the main elements in the controversy. At the same time, they both design a spatial work plan that combines the information collected by the technical experts of both Crowns with data found in Jesuit maps and the knowledge of "pilot"[8] Indians.

The significance of the map developed by Custódio de Sá e Faria is better understood if complemented by the reading of the Spanish map titled *Demostración del modo más fácil y exacto para poder averiguar el origen principal del Ibicuy y poder concluir la demarcación* [Demonstration of the easiest and most accurate way to find the main source of the Ibicuy River and conclude the demarcation] (Fig. 6.6). In a letter addressed to Ricardo Wall (Spain's State Minister), the Marquis of Valdelirios claimed that this map is a reproduction of the one drawn by Custódio de Sá e Faria, who had given it to him:

[8]In Portuguese, "pilots" are experts in a specific maritime area. This knowledge makes them indispensable individuals for local navigation. In the documentation about the parties, the term "pilot" is supposed to designate the Indians who were knowledgeable about the names of rivers, and their location, path, and depth. Indigenous peoples were then valued by the European boundary commissioners.

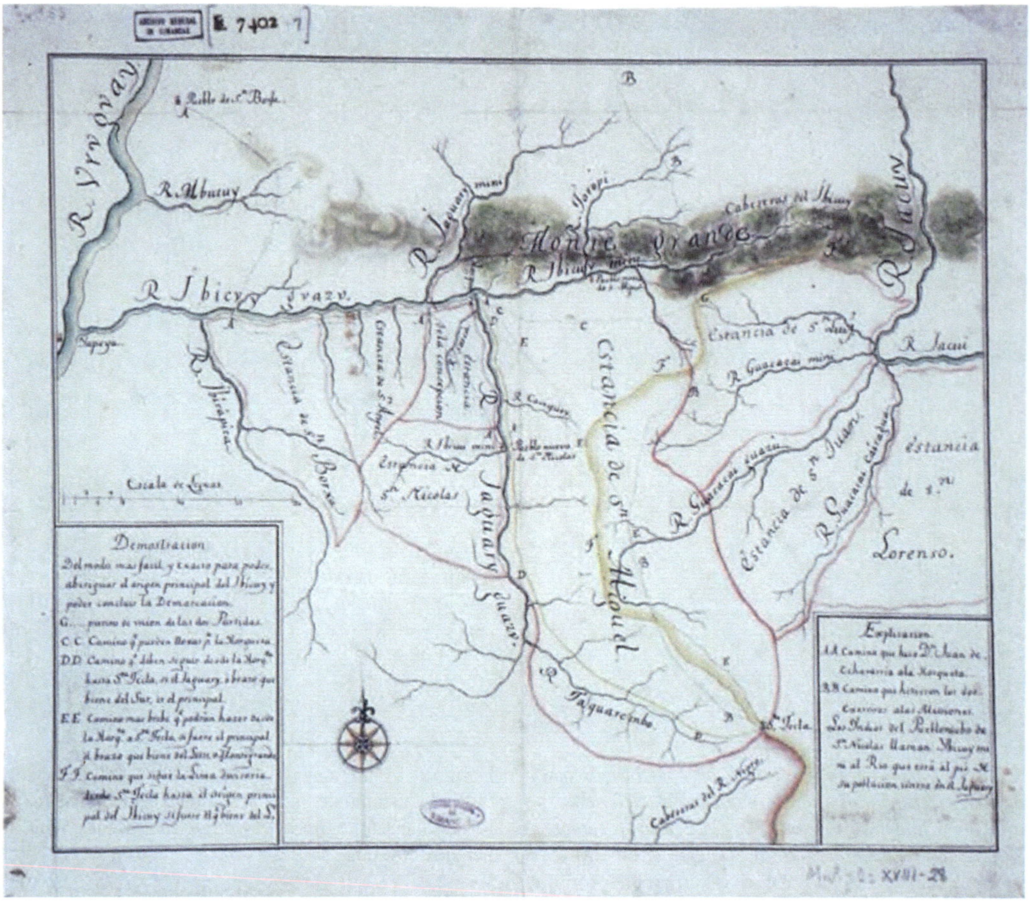

Fig. 6.6 *Demostración del modo más fácil y exacto para poder averiguar el origen principal del Ibicuy y poder concluir la demarcación* [Demonstration of the easiest and most accurate way to find the main source of the Ibicuy river and conclude the demarcation]. *Source* AGS. Mapas, Planos y Dibujos, 18, 028

> **It is a true copy** of the one he gave to me before leaving San Angel, developed by D. Joseph Custodio. This map proved to me the two Ibicuy rivers, which form the main river by this name, and which skirt the questioned land, which was full of Indians, as he stated, (…) so this Map recognizes two river arms, or Ibicuy rivers, resulting in the referred to questioned land, it now denies one of them, against these same testimonies. (Valdelirios to Wall, Campo de Santa Catalina, October 20, 1758, emphasis ours)

However, a comparison between *Demostración del modo más fácil y exacto…* [Demonstration of the easiest and most accurate way…] and the map by Custódio de Sa e Faria shows that this is not a “true copy.” The map drawn by Custódio de Sa e Faria seems to have been subject to a translation process (Latour 1999) inasmuch as the names of streams were adapted to conform Spanish interests. We will now identify the toponymic differences observed in the hydrography (Table 6.1):

The recognition of the areas where native peoples were settled should be added to this translation process as another difference between maps. Along these lines, the map called *Demostración del modo más fácil y exacto…* [Demonstration of the easiest and most accurate way…] gives more visibility to the lands occupied by native peoples to the south of the rivers that they called Ibicuy Guazú and Mini than the map developed by Custódio de Sa e Faria. While

Table 6.1 Toponymic differences between the Portuguese map (Fig. 6.5) and the Spanish map (Fig. 6.6) regarding the river courses included in the controversy

Portuguese names	Spanish names
Ybicuy	Ibicuy Guazú
Ybicuy Guaçu	Jaguary Guazú
Yaguari	JaguaryMini

Source Author's own elaboration based on a comparison of the cartography under analysis

on the Spanish map the areas occupied by native peoples are recognized as *estancias* (Estancias de San Miguel, Estancia de San Borxa, Estancia de San Angelo, Estancia de San Lorenzo, de San Luiz, Estancia de San Juan), in the context of which we can identify the settlements formed by indigenous migrants from the North (Pueblo Nuevo de San Nicolás, Pueblo Nuevo de San Miguel), the original map by Custódio de Sá e Faria only recognizes the new settlements. The indigenous peoples that inhabited the area before that time, the farming spaces and the newly arrived populations were thus made invisible.

We can also identify some differences in the maps' legends. Actually, the Spanish map *Demostración del modo más fácil y exacto…* [Demonstration of the easiest and most accurate way…] introduces itineraries to explore the area and solve the controversy. These are not incorporated to the original map drawn by Custódio de Sá e Fária.

Finally, the interpretation of the map *Demostración del modo más fácil y exacto…* [Demonstration of the easiest and most accurate way…] enabled us to recognize a set of tiers of significations that bring us closer to the entanglement formed by heterogeneous associations. First, we observed that the criteria at stake during the demarcation process cut across the streams whose names were in dispute (Which are the Ibicuy heads: the waters originating south-east and running diagonally or the waters forming a straight line with the river course in question?) Second, we identified a set of lines (in gray) and letters (G, C.C., D.D., E.E., F.F.) used to propose a meeting point and the itineraries for boundary commissioners to survey the land. Third, we located the *estancias* occupied by indigenous peoples (delimited with red lines), who, from the point of view of the Portuguese, hindered the progress of demarcation activities. These heterogeneous associations make temporalities from the past (indigenous occupations), present (the controversy), and future (exploration tasks and their implications in terms of territorial appropriation) configure spatialities that challenge the demarcation line on the Map of the Two Courts (in yellow) and that form part of the border reinvention process.

6.6 Final Remarks

The aim of this text was to approach the process of definition of the intercolonial border between Spain and Portugal in the Rio de la Plata region on the basis of the relation between the Treaty of Madrid, the Map of the Two Courts and field actions by boundary demarcation parties. We particularly focused on the party that worked in the area between Castillos Grandes and the confluence of the Ibicuy and Uruguay rivers. The vision of the Actor-Network Theory regarding the sociology of associations was helpful to understand the relations between heterogeneous non-human elements (like rules, observation and measurement instruments, the dynamic flow of streams, etc.) coupled with the actions of indigenous peoples and the practices of boundary commissioners during the process of border demarcation.

We have attached special importance to the analysis of water dynamics and the actions of native peoples inasmuch as they helped us recognize the significance of these agencies in the fabrication of a boundary that differed from the one originally proposed. Along these lines, it is our view that both humans and non-humans contributed to challenge the boundary political project, agreed on paper but detached from material conditions. Approaching the controversy about the identification of the Ibicuy river heads has not only allowed us to examine these agencies, but has also been useful for us to determine how they were incorporated to the argument positions presented by both sides.

Finally, the study of cartographic production unveiled the matter of the controversy and, in addition, through the processes of inscription and translation, led us to discern its role in conflict fabrication. The imaginary line proposed by the central governments through the letter of the Treaty of Madrid and the Map of the Two Courts was thus tensioned and challenged.

Archival Sources

AGS: Archivo General de Simancas.
BNB: Biblioteca Nacional do Brasil. Rio de Janeiro.
BNE: Biblioteca Nacional de España. Madrid.
Tratado firmado en Madrid, a 13 de enero de 1750, para determinar los límites de los Estados pertenecientes a las coronas de España y Portugal, en Asia y América. Alicante: Biblioteca Virtual Miguel de Cervantes, 2002. http://www.cervantesvirtual.com/obra/tratado-firmado-en-madrid-a-13-de-enero-de-1750-para-determinar-los-limites-de-los-estados-pertenecientes-a-las-coronas-de-espana-y-portugal-en-asia-y-america–0/ Accessed 29 Jan 2021.

Acknowledgements This work has been supported by the Spanish Ministry of Economy, Industry, and Competitiveness and the European Regional Development Fund (Research Project CSO2015-65301-P).

References

Cavalcante P (2006) Negócios de trapaça. Caminhos e descaminhos na América Portuguesa. Hucitec/ FAPESP, São Paulo
Erbig J Jr (2015) Imperial lines, indigenous lands: transforming territorialities of the Río de la Plata, 1680–1805. University of North Carolina at Chapel Hill Graduate School. Chapel Hill, NC: https://doi.org/10.17615/m6kt-tk94
Ferreira MC (2001) O tratado de Madrid e o Brasil Meridional. Os trabalhos demarcadores das Partidas do Sul e sua produção cartográfica (1794–1761). Comissão Nacional para as Comemorações dos Descobrimentos Portugueses, Lisboa
Gesteira H (2014) Um observatório astronômico nos confins da América Portuguesa-1750–1790. In: Gesteira H, Marinho P, Carolin LM (eds) Formas do Império. Ciência, tecnologia e política em Portugal e no Brasil. Séculos XVI ao XIX. Paz e Terra, São Paulo, pp 93–119
Gomes Castro C (2012) Demarcadores e Objetos em Campanha A Comissão Brasileira Demarcadora de Limites e a Fabricação de Fronteiras (1928–1947). Master dissertation. Postgraduate studies in social anthropology. National Museum. Federal University of Rio de Janeiro. http://objdig.ufrj.br/72/teses/778623.pdf. Accessed 20 Sept 2020
Herzog T (2015) Frontiers of possession. Spain and Portugal in Europe and the Americas. Cambridge, London
Latour B (2005) Reassembling the social—an introduction to actor-network-theory. Oxford University Press, Oxford
Latour B, Woolgar S (1979) Laboratory life: the social construction of scientific facts. Sage, Los Angeles
Latour B (1999) Pandora's hope. Essays on the reality of science studies. Cambridge University Press, Massachusetts
Martínez Martín C (2000) La nueva frontera Brasil-Río de la Plata en el Tratado de 1750: la demarcación del río Ibicui por la Primera Partida de Límites. In: Gutiérrez-Escudero A (ed) Ciencia, economía y política en Hispanoamérica colonial. Consejo Superior de Investigaciones Científicas, Sevilla, pp 433–450
Martín-Merás L (2007) Fondos cartográficos y documentales de la Comisión de Límites de Brasil en el siglo XVIII en el Museo Naval de Madrid. Terra Brasilis, 7, 8, 9. Dossier Cartografías Iberoamericanas. https://doi.org/10.4000/terrabrasilis.402
Moutoukias Z (1988) Contrabando, y control colonial en el siglo XVII. Buenos Aires, el Atlántico y el espacio peruano. CEAL, Buenos Aires
Possamai P (2014) Colonia de Sacramento. Vida cotidiana durante la ocupación portuguesa. Torres del Vigía, Montevideo
Quarleri L (2009) Rebelión y guerra en las fronteras del Plata. Guaraníes, jesuitas e imperios coloniales. Fondo de Cultura Económica, Buenos Aires
Sahlins P (1987) Natural frontiers revisited: France's boundaries since the seventeenth century. Am Hist Rev 95:1423–1451. https://doi.org/10.2307/2162692
Sahlins (1993) Fronteres i identitats: la formació d'Espanya i França a la Cerdanya, s. XVII-XIX. Vic: Eumo Editorial
Strandsbjerg J (2012) Cartopolitics, geopolitics and boundaries in the Arctic. Geopolitics 17(4):818–842. https://doi.org/10.1080/14650045.2012.660581
Tau Golin LC (2014) A guerra guaranítica. O levante indígena que desafiou Portugal e Espanha. Terceiro Nome, São Paulo
Tirado Serrano F, Domenech i Argemí M (2005) Asociaciones heterogéneas y actantes. El giro postsocial de la teoría del actor-red. AIBR. Revista de Antropología Iberoamericana, No. Extra 1: 1–26
Zusman P (2000) Tierras para el Rey. Tres fronteras y la construcción colonial del territorio del Río de la Plata (1750–1790). Ph.D. Thesis, Autonomous University of Barcelona

7 Rights of Use Versus Natural Boundary: The Demarcation of the Iraty Forest Massif (Western Pyrenees, Seventeenth–Nineteenth Centuries)

Jean-Yves Puyo

Abstract

I propose using the case of the Pyrenean border to analyse the development of European practices as regards the method leading to the establishment of a national demarcation line recognised by both parties. Situated on a mountainous massif, namely the Pyrenees range, this border has two major characteristics. On the one hand, its very recent establishment (dating from the mid-nineteenth century) led to the separation of local communities that had been united for centuries in kingdoms straddling both mountain slopes (the Kingdom of Aragon for the eastern Pyrenees, the Kingdom of Navarre for the western Pyrenees). On the other hand, the discussions between the two states, which took place in episodes straddling almost three centuries, were, depending on the border areas, marked by the abandonment of the recognition of "ancestral" rights of use of the local communities, in favour of the choice of so-called natural boundaries, a concept promoted in Europe by the French Revolution. Finally, this research makes a modest contribution to highlighting the complex relationships woven between the representatives of the core states and those of local communities, the rights of the latter being at times sacrificed on the altar of general interest, as in the case of the Iraty border massif for the inhabitants of the Pays de Cize.

7.1 Introduction

The southern and northern slopes of the Pyrenees range have the peculiarity of having been united for many centuries in territorial entities that disregard the mountains, such as the Trans-Pyrenean Kingdoms of Aragon and Navarre. Thus, the division between the two large hydrographic basins of the western Pyrenees did not take place until 1521, when Charles V definitively ejected from the Spanish peninsula the Albret dynasty, which had for several centuries reigned over a previously cross-border kingdom of Navarre, stretching from the Ebro to the Garonne; of its six former *merindades* (territorial divisions), namely Pamplona, Tudela, Estella, Sangüesa, Olite and Saint-Jean-Pied-de-Port, only the latter—also called Lower Navarre or French Navarre, situated on the northern slopes—remained independent of the kingdom of Spain.

In fact, it was not until the *Treaty of the Pyrenees* of 1659 that the entire east–west Pyrenean axis acted as a dividing line between the kingdoms of Spain and France. As far as the

J.-Y. Puyo (✉)
Laboratory TREE, UMR 6031, University of Pau and Pays de l'Adour, Pau, France
e-mail: jean-yves.puyo@univ-pau.fr

J. García-Álvarez and P. Puente-Lozano (eds.), *Beneath the Lines*, Historical Geography and Geosciences, https://doi.org/10.1007/978-3-030-96904-2_7

new border is concerned, Article 42 of the same treaty recommended that a general line should be based on "the Pyrenean Mountains [...] which had formerly divided the Gauls from the Spanish".[1] However, the methodology for determining the dividing line between the two sides was not set down in law, with only the application of "good faith" being stipulated. And, for more than a century, the question of its physical definition was not a priority for the two neighbouring powers; they therefore left the management of the border regions to the local communities, which the latter had already been looking after for some time.

In fact, the fourteenth century saw the first agreements called *faceries* being signed between the opposing valleys (Arvizu 1983; Desplat 1986); these were actual treaties of alliance and peace governing the field and woodland pastoral use of the border areas. Some of these texts were later used as a basis for discussions held at the state level. For example, the *Royal Capitulations* of 25 September 1614 (or the Treaty of San Lorenzo), designed to maintain a state of peace between the border communities of Navarre, did not address the question of the demarcation of the two kingdoms; they limited themselves to specifying the rights of use and their perimeters of application, here in the case of the Aldudes valley (western Pyrenees).

However, from the very beginning of the eighteenth century, the increase in border conflicts in these same areas (Arvizu 1992) eventually forced the two states to tackle the question of their shared border, which was still unresolved. This was a thorny problem that raised numerous challenges, including methodological ones. For example, what elements should a border both parties considered "strong" subsequently be based on? Should its establishment take into account rights of use or, on the other hand, be based on natural boundaries in accordance, for example, with the dividing line of the waters in the large catchment areas here in the Pyrenees?

It is this line of questioning that we propose to study, through discussion of the vast Iraty border forest massif.[2] In fact, as one of the main stumbling blocks in the border demarcation in the western Pyrenees, the French decision to leave large areas of forest to Spain was not without consequence for the success of the bipartite negotiations that led, in the mid-nineteenth century, to the determination of a Pyrenean border line that has not been contested since.

7.2 The Iraty *Faceries* (Fifteenth–Eighteenth Centuries)

Because of the ease of accessibility during a large part of the year, the development of the intra-Pyrenean space by traditional local societies has given rise to a multitude of secular neighbourhood conflicts between shepherds grouped, by valley, into "autonomous republics" (Desplat 1986). Thus, from the beginning of the fourteenth century, the first *facerie*-type agreements signed between the opposing valleys determined the demarcation of ownerships, the use of pastures, water and wood resources, access roads and the management of hunting. According to French geographer Henri Cavaillés: "[...] There was not a rock, a blade of grass or a thin trickle of water on the most inaccessible peaks [of the Pyrenees] that was not attributed to one or other of the valleys, and whose use was not specifically regulated" (1910, p. 17). At the same time, a whole system of penalties was provided for in the event of infringements, such as the possibility of seizing offending livestock (known as the right of redemption ("droit de *pignore*") or the right of confiscation ("droit de *carnale*") of livestock), as well as compensation in the event of brawls, with the application of a progressive scale according to the seriousness of the damage recorded. Finally, public agents appointed by each valley were responsible for ensuring the proper implementation of these pacts. Thanks to these local

[1]Article 42 of the Treaty of the Pyrenees, extract from the *Traitez de Paix, de trêve, de neutralité, de confédération, d'alliance et de commerce*, etc. (Léonard 1693, p. 19).

[2]It covers almost 17,200 ha, almost 4/5ths of which are nowadays located in Spanish territory (Brocas and Legaz 2004).

cross-border agreements, again according to Henri Cavaillés, "[…] conflicts became much rarer and, above all, less serious fairly early on". (ibid.)

Some treaties even established freedom of trade between the slopes, as well as a certain neutrality in the event of national armed conflicts, with the communities undertaking to warn signatory parties in the event of an assembly of belligerent troops on the border; this type of provision, still in use in the mid-seventeenth century, then fell into disuse, the end of the Wars of Succession for the Spanish throne (1720) marking the beginning of a long period of peace locally.[3]

In Lower Navarre, the use and ownership of the Iraty forest massif—the subject of our study—was established at the end of the Middle Ages by a series of treaties of good neighbourliness between the inhabitants of the Salazar valley (southern slope, Spanish) and those of the Pays de Cize (northern slope, French), the oldest of which dates back to 1445, well before the incorporation of Upper Navarre into the kingdom of Spain. Subsequently, a new treaty signed in 1507 established a kind of jurisprudence for almost three centuries, which was renewed as it stood until 1759 for a period of nine years.[4] That was when the problems began.

On the part of the French, this text of 1507 is considered as allocating the ownership of the massif to the Pays de Cize, a point of stalemate between the two parties later on: "[…] the boundary between the two countries is determined by the main waters of the Irati from the markers that mark the boundary with the Soule to the river Iasola [Urchuria], and then following the waters of Irati to the boundary markers of the Pays d'Ahescoa [Aezkoa]".[5] It also regulates the use of the grazing areas located within its borders, whether these are high altitude pastures (located above the forest level) or old forest areas that were subsequently deforested. These pastoral areas are subject to the so-called right of *compascuity*, i.e. the right to graze herds belonging to multiple communities of inhabitants, in this case located on either side of a border. The text further stipulates that herds from both valleys may graze on a neighbour's land from sunrise to sunset, returning to their respective territories in the evening; and, in the event of a breach of this regulation, seizure of the offending herds may only be carried out "[…] after giving three days' notice".[6] Finally, an important point often at the root of disputes, both communities are prohibited from introducing outsider-owned livestock onto these pastures, as well as from holding tree sales. And it was precisely this issue of the exploitation of forest resources that led to the greatest of the stumbling blocks.

For the authorities of the kingdom of France, the Iraty forest was long considered to be "the end of the world" with no great economic value, located far from any habitation, with no passable roads and, above all, no navigable rivers enabling the exploited timber to be transported by floating. The first major French experiment to develop the forest massif dates back to 1629, when private individuals in the region's capital, Saint-Jean-Pied-de-Port, tried to remove fir trees suitable for making naval masts from the forest. But of the 7000 ft bought from the Pays de Cize syndicate, only about sixty could be cut down and taken to the river Nive to be floated down to the royal arsenal in Bayonne. Attempts to supply wood to the French Royal Navy seem to have continued without much success, but, paradoxically, in the middle of the eighteenth century, more successful exploitation of these same firs, this time on behalf of the Spanish Navy, would singularly complicate the local situation, after three centuries of "entente cordiale".

This new interest on the part of the kingdom of Spain in Iraty's forest resources stems from a combination of the following two facts. On the one hand, according to French accounts of the

[3]For more information, see among others the study by Colonel Bernard Druenne (1986).

[4]Treaties of 1556, 1568, 1569, 1609, 1618, 1669, 1717 and 1751.

[5]Gleizes, *Avis du commissaire français sur les discussions entre le pays de Cize et la vallée de Salazar*, June 1829, 1 J art. 105-3, ADPA.

[6]Ibid.

time, the woods of the neighbouring forest of Abodi—located on the left bank of the Rio Irati—were by then exhausted by past over-exploitation; this fact would have led the Upper Navarre (Spanish) loggers to cross the rio, and therefore the border, to undertake cutting, this time in the woods bordering the Iraty massif. On the other hand, and as will be illustrated later, the same Rio Irati, whose course is part of the Ebro river basin, considerably facilitated the transportation of the exploited timber, with the neighbouring villages in the Upper Navarre region providing the necessary manpower.

Thus, in 1766, barely seven years after the last renewal of the faceries, the *alcade* of the Salazar valley succeeded in selling the right to exploit the Iraty forest—where, according to a French chronicler, the wood proved to be of better quality than in the Abodi forest—to the Spanish crown.[7] This work caused such a stir on the Cize side that Spanish troops were sent to protect the loggers at work, before the French government was temporarily satisfied through the suspension of logging. However, this resumed in 1781, to the great despair of the local French authorities.[8]

7.3 The First Local Attempts at Border Demarcation (1784–1792)

As border incidents multiplied, the two rival powers set up a bipartite commission in 1784 with the task of carrying out precise cartography of the Franco-Spanish border zone, from the Atlantic to the Mediterranean, and to jointly propose a demarcation between the two kingdoms that would be acceptable to both parties. Known as Caro-Ornano, it constitutes an important moment of cooperation between the two nations (Arvizu 2001; Capdevila 2009; García-Álvarez and Puyo 2015, 2019; Puyo and García-Álvarez 2016; García-Álvarez and Puente-Lozano 2017a).

Paradoxically, during this commission's work and while the two sides were in constant contact, there was an increase in border incidents in the Iraty sector, including, in September 1785, the Iraty forest suffering a very significant arson attack lasting three consecutive days, the fire having been set in seven different places in the massif. While the Spanish frontiersmen set about fighting the fire, the Cize frontiersmen were strongly suspected of being the perpetrators. Feelings were heating up to a very dangerous level at the same time,[9] and Count d'Ornano had all the French shepherds living in the Iraty forest arrested. They were not released until the investigation carried out by the bipartite commission showed that the fire was the work of Spanish smugglers.[10] In fact, military action had only just been avoided. However, the incidents did not go away, such as the official French protest by d'Ornano in May 1785, after Spanish woodcutters had carried out "[...] a 19-mast cut in the part of the forest that belongs to us".[11] Also, six

[7]"*Projet de mémoire pour établir les droits des communautés de la Vallée de Cize à la propriété de la forêt d'Irati*", an undated 7-page manuscript (presumably *circa* 1858–1860), signed *Le conservateur des Forêts (The Curator of Forests)*, ADPA, coll. 7 M 165.

[8]"[...] in recent years, the Spanish have cut down an immense quantity of trees for the benefit of their navy and to the detriment of that of the said Lord King, as recently noted by a naval commissioner on the orders of the Secretary of State of this Department". Parliament of Navarre, *Remontrances sur la démarcation des limites entre la France et l'Espagne sur les limites frontières*, 29th August 1786, ADPA, E 2321.

[9]As Caro had then brought troops close to the border, d'Ornano, for his part, was content to send a brigade of the marshalcy to investigate. Letter from Count d'Ornano to the Count of Vergennes, French Minister of Foreign Affairs, Saint-Jean-Pied-de-Port, 8 September 1785. Limits Series, Sub-Series Spain, vol. 18, ADMAE, La Courneuve site.

[10]45 years later this episode gave rise to a new investigation (!), on the occasion of the work of the Gleizes-Bayona bipartite commission (cf. Puyo 2018). Thus, the Pyrénées-Atlantiques departmental archives contain the Spanish transcription—and their translation into French—of 23 testimonies of inhabitants from the Salazar valley who knew or heard about the fire. 1 J art.105-1, ADPA.

[11]Letter from Count d'Ornano to the Ministry of Foreign Affairs, Iraty, 04 October 1785. Limits Series, Sub-Series Spain, vol. 18, ADMAE, La Courneuve site. In the same letter, d'Ornano reports that he had asked Pamplona for the return of a herd of cattle illegally seized in the Iraty forest.

years later, there is mention of a collection of firewood being carried out by the people of Cize, which was repulsed by Spanish troops assisted by a large number of peasants from the Salazar valley, amongst others.

Of the achievements of the Caro-Ornano Commission—and right up to our recent work dedicated to its scientific and cartographic legacy (Puyo and García-Álvarez 2016; García-Álvarez and Puyo 2019)—history had recorded relatively little, namely the only episode (or almost the only one) relating to the signing of the Treaty of Elizondo on 27 August 1785. This was intended to provide a lasting solution for the neighbouring Aldudes valley to the then very sensitive issue of border demarcation and field and woodland pastoral use (Arvizu 1992; Capdevila 2009).[12] From then on, the border between the Aldudes and Valcarlos valleys was established through the installation of a little more than 160 markers along a line of a little more than 44 kms, i.e. one marker installed every 275 m.[13]. However, as far as Iraty is concerned, the episode involving the Caro-Ornano commission was not without consequences.

Firstly, it gave rise to a major joint topographical survey, resulting in the publication of two maps of the Iraty Massif (among them, the one reproduced in Fig. 7.1), at a scale of three leagues per 100 toises—approximately 1:29,000 —which are included in the Spanish and French collections (García-Álvarez and Puyo 2019).[14]

Secondly, the joint recognition of the border confines by Commissioners Caro and d'Ornano enabled the representatives of the communities concerned to reaffirm their positions regarding their rights over the disputed massif. As Daniel Norman explains, this approach corresponded to the current territorial model, based on a pyramid-like superimposition of rights attached to the place in question (Nordman 1998). Thus, on 7 June 1788, the deputies of Cize, Saint-Jean-Pied-de-Port and Aezcoa were summoned to boundary marker 197 in Iriburietaco-Lepoa and were given a public reading of "[…] documents that each of them could produce in justification of their rights, […] aloud, in the common Basque language, […] and in the Spanish and French idioms".[15] All the parties then recognised "[…] as the true dividing line between their properties and possessions, that specified in the aforementioned arbitral award of 13 August 1556".[16]

Ten days later, a new meeting of all the parties concerned (deputies from Cize, Saint-Jean-Pied-de-Port and Salazar), this time at the place known as Idaybea, resulted in the same successful outcome for the two investigating commissioners, namely the recognition of the former treaties.[17] But difficulties arose the very next day (17 June 1788) with the surprise arrival of a delegation from Aezcoa (uninvited, no less) who wanted to submit a text presented by them as newly discovered and dating from 1 October 1423. This led to the formation of a common cross-border stance, uniting the representatives of Cize, Saint-Jean-Pied-de-Port and Salazar, who firmly refuted this text "[…] insofar as this might oppose their rights of ownership and possession".[18] In fact, they all pointed out that they had never heard of it until that day! As a result of this disagreement between the parties, and contrary to the work of 1785 in the neighbouring Aldudes valley, the recognition of the Iraty Massif made it

[12]*Definitive boundary treaty between France and Spain, to establish a dividing line between the Aldudes or Quint-Royal and Val-Carlos in Navarre, and to determine the boundaries of the two states.* This treaty was subsequently ratified by Louis XVI on 5 March 1786 and by Carlos III on 21 March 1786 (Arvizu 2001, p.23).

[13]Letter from Count d'Ornano, Saint-Jean-Pied-de-Port, 8 September 1785; sub-series Spain, vol. 18, ADMAE.

[14]*Plano topográfico de los Montes de Yrati, término común de Aézcoa y sus inmediaciones levantado de orden del Sr. Dn. Ventura Caro Mariscal de Campo de los Res. Cxtos. de S.M.C. / Por los Ingenieros Dn. Juan Casanoba y Dn. Josef Martinez*, at a scale of 1: 29,000 (ACEG-GCE, Ar. H-T.5-C.8-236); as well as a *Plano ideal del bosque Yrati y de la Mojonera entre los dos valles de Aézcoa y de Cisa según el reconocimiento que hicieron los ingenieros destinados a la demarcación de límites bajo las órdenes del mariscal de campo Dn Ventura Caro* (BNF, Département Cartes et plans, GE C-3075).

[15]1 J art.105-2, ADPA.

[16]Ibid.

[17]Arbitral award of 7 September 1507, as well as the agreements of 5 September 1669 and 8 August 1759: "*They unanimously replied that they recognised them as genuine in all their content*". Ibid.

[18]Ibid.

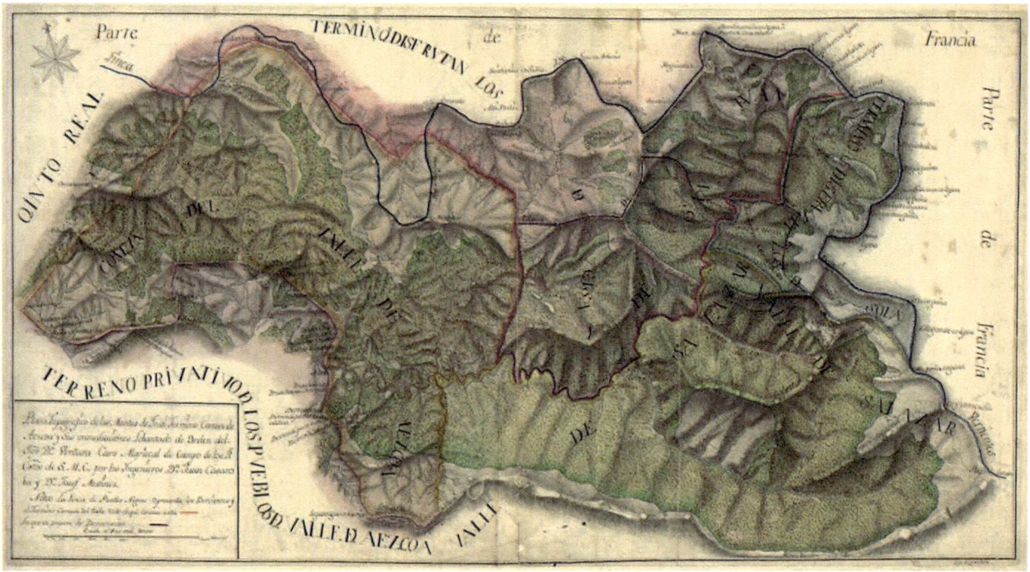

Fig. 7.1 *Plano topográfico de los Montes de Yrati,* by Juan Casanoba & Josef Martinez (*circa* 1790, scale 1:29,000). *Source* ACEGCGE, Ar. H-T.5-C.8–236

impossible for a demarcation line to be established and subsequently marked out between the two states. The failure was obvious, as the valley of Aezcoa had put a cat among the pigeons by claiming "new" rights to the same area.

7.4 Respect for Rights of Use Along the So-Called Natural Border

As Jacques Ancel, one of the great pioneers of the geographical study of borders (García-Álvarez 2019), pointed out in 1938, the French Revolution was to promote a new way of conceiving state borders in Europe, based on so-called natural boundaries. So it was that, on 27 November 1792, in a speech to the National Convention, in which he was attempting to justify the planned integration of Savoy into France, Abbé Grégoire called for "the archives of Nature to be consulted [while respecting] the boundary markers set for the French Republic by the hands of nature": "France is a whole that is self-sufficient in itself, since everywhere nature has given it barriers that exempt it from expanding, so that our interests are in agreement with our principles" (Mavidal and Laurent 1898, p. 610).

This speech caused a sensation, raising cheers from the members of the Convention.[19] It should be noted that this project had previously given rise to a consultation with the "Allobroges people".[20] A few months later, when the question of a meeting in the French Republic of Belgium arose, the Belgian people were this time, to paraphrase Jacques Ancel, neither consulted nor consenting (Ancel 1938, p. 75). Before the Convention, Danton defended this annexation project out of the concern "of one great people [the French nation] to bring freedom to another people" (Fribour 1910, p. 268). And to his opponents, who wondered about the limits of this territorial extension of the French Republic, he advocated, following the example of Abbé Grégoire, "boundaries marked by nature": "We will

[19]"Nature had already decreed the physical and moral unity of our common territories, and we have just obeyed it; and it will not be the last tribute that the Convention will boast of paying to the inspirations of nature". Hérault de Séchelles quoted by Mavidal and Laurent (1898, p. 617).

[20]On 22 October 1792, the assembly of representatives of the 655 Savoyard communes voted overwhelmingly in favour of this integration with France, with 580 votes in favour of the project. Quoted by Abbé Grégoire (Ibid., p. 612).

reach them at their four points—at the Ocean, the Rhine, the Alps, and the Pyrenees; no power can stop us" (ibid.).[21] For Albert Sorel, in his speech of 31 January 1793, Danton expressed the predominant thinking among the Convention members "[…] by showing natural boundaries as being the necessary expression of the republican war" (Sorel 1891, p. 345).

In fact, as André Zeller points out, the use of so-called natural borders is a very old idea which, in the French case, had "been on its way" for a long period of time, the first border projects on the Rhine being as early as the fifteenth century, for example (Zeller 1933). However, for the same author, it is completely wrong and abusive to see the French Revolution as the simple promoter (and continuer) of a concept that should be owed to the Ancien Régime.[22] In any case, the recourse to natural boundaries for the demarcation of national borders was a lasting integration of the republican "software" and not without potential abuses: "Manipulated without qualification and used indiscriminately, the natural border could become an instrument of annexation and, subsequently, of oppression" (Guichonnet and Raffestin 1974, p. 19). Subsequently, the idea of using natural borders for the territorial limits of the French state was interpreted in the nineteenth century as a symbol of national unity reconciling the principles of the Ancien Régime with those of the First Republic, before being generally conveyed to a wider public through history and geography textbooks (Sahlins 1990). At the same time, as Jacobo García-Álvarez and Paloma Puente-Lozano (2017b) point out, from the mid-eighteenth century onwards, the neighbouring kingdom of Spain resorted to natural borders to draw new fixed and permanent boundaries between their South American possessions and those belonging to Portugal (Treaty of Madrid, 1750, and Treaty of San Idelfonso, 1777).[23]

Subsequently, the Treaty of Basel (22 July 1795), which ended a period of war between France and Spain, undeniably marked the success of this plan (Foucher 1986). During the preliminary discussions leading up to the drafting of this treaty, the question of the "old Pyrenean boundary matter" (Fain 1828, p. 229) was raised, and, in an attempt to settle the disputes in progress, the two plenipotentiaries, François Barthélemy and Domingo d'Yriarte, decided to take "the law of the slopes" ("la loi des versans") (ibid.) as their invariable basis, whatever the cost to both parties.[24] And so, in Article VII, this treaty recommended that the Franco-Spanish border demarcation should be based on the delineation of the major hydrographic basins: "Commissioners will soon be appointed on both sides to draw up a boundary treaty between the two powers. As far as possible, with regard to the lands that were in dispute before the present war, they will take as a basis for this treaty the ridges of the mountains that form the slopes of the waters of France and Spain".[25]

However, the planned meeting of a bipartite commission to establish the new border on these natural boundaries did not go ahead, fuelling local discontent, particularly in the Iraty forest where French foresters continued to cut down stands on the southern slopes. In 1803, for example, a certain Etcheverry, a merchant from Bayonne, obtained a concession to extract 300 cartloads of oars, provoking strong protests from the representatives of the Salazar valley to the Viceroy of Navarre. In April 1804, the Viceroy

[21]It should be noted that there are several formulations of this same bullish talk, according to the media that reported it at the time (the *Journal des Débats*, the *Moniteur*, etc.), but they all share the same general spirit: the very young French Republic is called upon to extend itself out to its so-called natural boundaries.

[22]"The policy of natural borders was neither monarchical nor national under the Ancien Régime, but was only the work of a few isolated individuals, a few chimerical minds" (Sorel 1891, p. 331).

[23]Article 5 of this treaty stipulates "[that] the Portuguese shall not cross the Tahim river, nor the Spaniards the Chui and Saint-Michel rivers" (Kock 1817, p. 227).

[24]"If this principle is rigorously applied, French Cerdagne could be lost to the Republic. After various drafting attempts, the Committee's plenipotentiary decided to consent to the project, with the ulterior motive that the same principle can give us the Aran Valley in compensation" (Fain 1828, p. 230).

[25]Ibid., p. 415.

sent an official complaint to Paris[26]; in response, the French Minister of Foreign Affairs reminded the Prefect of the Lower Pyrenees of his previous instructions (dated 22 January 1804), prohibiting tree cutting in Iraty until the demarcation had been carried out. In fact, the Curator of Water and Forests, charged with enforcing this decision locally, should never have authorised this sale which, in addition to the diplomatic incident, caused serious local tensions likely to deteriorate dangerously.[27] Subsequently, the enforced abdication of Charles IV, followed by the disastrous invasion of Spain by Napoleon's troops, once again postponed the resolution of the thorny problem of the border line until the Treaty of Paris of 1816, which put an end to the Napoleonic Wars: this treaty recognised France's borders on 1 January 1792, effectively rendering the recommendations of the Treaty of Basel null and void…

But it was a surprise to us to discover that the local communities of these Pyrenean confines were very aware of the content of these latest international treaties! Thus, in 1826, the representatives of the Salazar valley officially defended the desire for a border demarcation based on the recommendations of the Treaty of Basel, namely the study of natural boundaries (mountain ridges, water slopes, etc.), to the fury of the inhabitants of the Cize valley.[28] The representatives of the latter replied that this treaty had never been executed and, moreover, that "[…] under the Treaty of Paris of 30 March 1814, things [had] been restored to the state they had been in before".[29] To which their Salazar opponents replied unequivocally that, even though the Treaty of Basel had been abrogated by the Treaty of Paris—"[…] of which they had only become aware at that time"[30]—the treaty signed at the so-called Pheasant Island Conference (or the Treaty of the Pyrenees of 1659) had established the principle adopted by the Treaty of Basel, namely "to set the demarcation by the mountain ridges, and that they should stick to it; and that the other titles are no longer of any consideration".[31] And even if this last statement proves to be erroneous (see our introduction), the fact remains that these exchanges leave us with an image of a local elite that is highly educated in the content of international agreements that directly concerns it!

As incidents increased in frequency during the 1820s in the border area of the western Pyrenees, the two states agreed—not without difficulty—to set up a new bipartite commission in 1829, named Gleizes-Bayona (Puyo 2018). One of the most important points to be noted about this temporary episode (1829–1831) and the issue of the Iraty massif is the following: for the first time, the French investigating commissioner,

[26]Letter from the Viceroy dated 18 April 1804: "We see there that it was citizen Ferié, brigadier [of Waters and Forests] residing in Saint-Jean-Pied-de-Port, who allowed citizen Etcheverry of Bayonne to proceed with the purchase by him of a certain quantity of oars to be taken from Irati forest, taking advantage of what you had approved; that he has had the more than reprehensible flippancy of authorising the forcible repulsion of those who would try to oppose the execution of his illegal permission, thus causing serious accidents and misfortunes which the restraint of the Spanish forestry officers seems to have prevented". Letter of 18 Nivôse, year 12, ADPA, coll. 7 M 165.

[27]The general guard of Waters and Forests in charge of the massif would have "[even] had the more than reprehensible flippancy of authorising the use of force to repel those who would try to oppose the execution of his illegal permission and thus cause accidents and misfortunes that the restraint of the Spanish forestry officers seems to have prevented". Ibid.

[28]"As a result, the entire Irabié district, those of Igurguy and Lissardoya and the Iraty forest belonged to them". *Report of 17 August 1826 made to the administrative commission for the communal revenues of the former Pays de Cize by Messrs. de Lafaurie, Laurent, Etchepperestou (Justice of the Peace), Salabery (royal notary) and Irigoin, commissioner appointed by the deliberation of 24 June last on the disputes between the said Pays de Cize and the Salazar valley, concerning the movement of livestock on the respective bordering mountains;* 1 J art. 105-6, ADPA.

[29]The Treaty of Paris agreed a return to the frontier boundaries of 1 January 1792, with a joint commission being made responsible for setting the final demarcation (Article 3).

[30]*Report of 17 August 1826 to the administrative commission for the communal revenues of the aforementioned Pays de Cize* etc., op. cit.

[31]Ibid.

Colonel Gleizes, mentioned the possible transfer of the Iraty forest to Spain, in return for the payment of compensation to the Pays de Cize, in the event that the future border would represent both a line separating the two states politically as well as, "[…] at the same time, the absolute separation of the ownership of the subjects of the two crowns",[32] which was not the case at the time, at least for those in favour of recognition of the secular rights of the Pays de Cize.

7.5 From a Dispute Over an Obscure Logging Operation in Iraty to the Establishment of the Pyrenean Border (1852–1862)

The demarcation of the new border was once again pushed back as a result of political instability in both kingdoms. In Iraty in 1838, the French government undertook new operations to delineate the forest massif, but curiously did so unilaterally, as was revealed by an inspector from the French forestry corps: "This work will remain incomplete since, in view of the state of revolution in which Spain finds itself [and in particular Navarre, a Carlist stronghold], it is impossible to imagine having the involvement of a delegate from that country; but until now, no difficulty has ever arisen over the forest boundary from this Kingdom's perspective [sic] and at least the demarcation will apply to the rest of the perimeter".[33] This French demarcation, which gave rise to a national decree dated 11 August 1839, was never reflected locally by a prefectoral decree, so that in 1846 French foresters were still wondering about the official delimitation of the massif under their management[34]… And it was precisely this umpteenth dispute regarding a French logging operation in Iraty, together with the launch of new discussions between the two States, which was to give impetus to the process, leading to the establishment of the border for the whole Pyrenean massif in less than 10 years—an operation which, as has already been pointed out, had been pending for almost three centuries!

On 21 September 1852, the syndicate of Cize awarded the cutting of beech trees intended for making oars to a certain Dominique Landerretche, living in Saint-Jean-Pied-de-Port and acting on behalf of the Maison Bacqué in Bayonne. And the following month, the general guard of Waters and Forests in charge of the Iraty-Cize sector reported to his superior the arrival in Saint-Jean-Pied-de-Port of Spanish forest rangers, strongly protesting against the new cutting by this Landerretche. They even threatened to seize the woodcutters "under the pretext of a provisional agreement",[35] which the guard said he did not know existed: "Should I help the successful bidder to carry out his contract, with the assistance of the troop camped at Lecumberry, or should I refer to this agreement, which states that all the woods in the region of the border are to be respected without anyone being allowed to cut them down?".[36]

And indeed a letter from the Director General of Forests, dated 6 October 1852, informed the Curator of Pau of the very recent launch of political negotiations with the Spanish government concerning, once again, the demarcation of the Pyrenees border. In the meantime, the French Minister of Foreign Affairs asked for the Bacqué award to be postponed and, on the other hand, for "any kind of sale that might have been authorised in the woods in question [Iraty communal woods] to be postponed […] until further

[32]Colonel Gleizes, *Avis du commissaire français sur les discussions entre le pays de Cize et la vallée de Salazar*, Arnéguy, 23 October 1829; 1 J art.105-3, ADPA.

[33]26 November 1838, ADPA, coll. 7 M 861.

[34]"I don't think that the demarcation has been officially brought to the attention of the Spanish residents or that they have been asked to be involved in this operation, and perhaps it has not been approved either. Moreover, the Inspectorate's records say nothing about this". Letter from the Forest Inspector in charge of the Iraty massif to his superior in Pau, 22 August 1853, ADPA, coll. 7 M 861.

[35]Marin, letter to the inspector of Waters and Forests (illegible), Saint-Jean-Pied-de-Port, 20 October 1852; ADPA, coll. 7 M 861.

[36]Ibid.

notice".[37] And even though in January 1853 the two States had agreed to declare a ban on logging in the disputed areas of Iraty, at the local level neither party was ready to comply with that status quo.

For example, in June 1853, the French forest rangers seized 400 pieces or "large beams" in the floating port of Irabie, which had come from the disputed area, thereby undermining all theoretical agreements... In fact, the size of the forest massif and its remoteness from the French villages made close surveillance almost impossible for a large part of the year; the forest rangers could only guarantee a permanent presence there during the summer months. Moreover, the Navarrese offenders had only to cross the Rio Irati, which was then the border, to escape being fined, and once the French authorities had left, they could once again resume their logging work with impunity. And this is what happened in this episode dating from June 1853, described above: the guards and the military troop left for their camp 14 kms away (in the village of Lecumberry), the Spanish lumberjacks crossed the Irati river again and continued to evacuate the woods by raft, to the great fury of the Cize syndicate: "And this is how the Spaniards observed the status quo agreement".[38]

At the same time, as far as the French were concerned, the Bacqué cutting finally stopped, albeit rather belatedly, on 17 July 1853.[39] The argument made to explain this continued exploitation was based on the fact that it was carried out in a forested area (Oraatéco-Thomba) that had never been the subject of disputes with the Salazar valley over the centuries.[40] In any case, in order to protect the loggers at work, two companies of the 66th Line Regiment (infantry) were positioned on the border for a time, withdrawing further inland once the Bacqué cutting had been interrupted in order to avoid bloody clashes with the Spaniards. And shortly after the departure of the French soldiers, armed Spaniards set fire to the barracks that the troop had built and then abandoned, "[...] beating up an unfortunate French worker returning from Larrau, whom they had met there, without our troop having made the slightest movement aimed at preventing this in vain".[41]

This withdrawal of French troops in fact corresponded with a significant advance in the negotiations between the two states which, at the end of July 1853, agreed to define two lines of occupation in Iraty with troops "[in order] to prevent deplorable brawls and to prepare the way for the new negotiations that [were about] to begin".[42] However, even if this was accepted in principle, it took several months of "bargaining" before the bipartisan commission convened for this purpose finally agreed on the route of the two notorious demarcation lines.[43] This was completed on 19 November 1853, when the international convention dedicated to the neutralisation of the Iraty-Cize forest was signed.[44] Well aware that this decision was bound to raise many protests on the French side—and they were

[37]Letter from the Director General of Forests to the Curator in Pau, Paris, 6 October 1852; ibid.

[38]Report by M. Salabery, Pays de Cize syndicate (Salabery) to the sub-prefect of Mauléon, St Jean-Pied-de-Port, 26 July 1853; ADPA, coll. 7 M 861.

[39]By that date, only 125 trees had been harvested out of the 900 awarded. A lengthy legal procedure ensued, with the Maison Bacqué asking both the State and the Cize syndicate to pay compensation for the damage suffered, as the 775 missing trees could never be delivered. An agreement between the parties finally came about in September 1859 with the payment of 1000 francs to Mr Bacqué and Mr Landerretche, "wishing to remove any cause for litigation". ADPA, coll. 7 M 861.

[40]It should also be noted that it will "remain French" after the final border demarcation.

[41]Report by M. Salabery to the sub-prefect of Mauléon, op. cit.

[42]Letter from Drouyn de Lhuyn, Minister of Foreign Affairs, to the Minister of Finance, who was then in charge of the French Forestry Corps, Paris, 23 July 1853; ADPA, coll. 7 M 861.

[43]Hence, for example, the forest district of Oraatéco-Thomba incorporated the neutralised zone.

[44]General Callier, Baron Gros, letter to the Prefect of the Lower Pyrenees reporting on the agreement of 19 November 1853, Bayonne, 22 November 1853; ADPA, coll. 7 M 861.

not wrong on this point[45]—the French officials of the commission stressed the fact that this neutralisation was only temporary, and even called for the cooperation of the local population: "The inhabitants of our borders will easily understand all the strength that their restraint will give to their rights and, by their conduct, they will help us make their interests prevail, these being currently in our hands".[46] But they were unsuccessful…

And seven months later, in May 1854, a new agreement signed by the Governor of the Province of Navarre and the Prefect of the Lower Pyrenees set up a joint brigade consisting of three French forest rangers and the same number of Spaniards, responsible for monitoring the neutralised area by way of joint patrols.

7.6 In Conclusion: Where the Pays de Cize is Stripped of Most of the Iraty Massif

Three years later, the work of the Franco-Spanish commission that met in its early days to settle the Iraty problem led to the *Treaty of Limits*, signed in Bayonne on 2 December 1856, followed by two other agreements designed to settle the whole Pyrenean border, from the Atlantic to the Mediterranean.[47] In this first treaty of 1856, the Upper Navarrese abandoned their claims to the whole of the Iraty hydrographic basin, which would have led to the retrocession of the entire forest massif to them; however, in return, they were given most of the same southern slope of Iraty as far as the Abodi forest (Fig. 7.2).

Henceforth, the course of the Rio Irati no longer represented the Franco-Spanish border, the latter passing further north, along the so-called Ahuntzbide ridge, to the great displeasure of the Cize syndicate. After having protested in vain against a forced and (in its eyes) unjustified transfer, the syndicate succeeded in getting the French state to admit the dispossession of which it considered itself to have been a victim, and which was not contested by the Ministry of Foreign Affairs at the time: "The Cize valley had full and complete ownership of [the Iraty massif]".[48] As Colonel Gleizes had stated a few years earlier, in the event of a transfer of ownership to Spain, the State would have to compensate the citizens of Cize who had been dispossessed. And so it happened that a commission was set up in 1858 to determine the compensation to be paid; thus, the Iraty pastures and the 1263 ha of woodland—mainly in the form of high forest—that had been ceded were valued at nearly 154,000 francs, a sum which the French state paid to the Pays de Cize syndicate in October 1860.

Following the example of local public opinion, the final result of the Franco-Spanish negotiations shocked many people on the French side. Take, for example, this acerbic opinion, dating from 1872, by the geographer Onésime Reclus: "From the Mediterranean to the Ocean, it would have been easy to separate France from Spain by a reasonable route; it would have been sufficient to follow the dividing line of the waters. This was not done everywhere […] We have gained 5200 ha on the upper Iraty, but have lost 16,800 at the origins of the Nivelle and the Nives. So we do not have the advantage in the exchange of territory that these counter-intuitive limits entail. The diplomats of both nations have been at work numerous times over the last 250 years to draw these borders, but neither France nor Spain have anything to be proud of" (Reclus 1872, p. 610).

[45]"All our neighbours are triumphant and boast as if they have won a decisive battle […]. Every day they violate our territory by running arms there […] Ultimately, they act as if they are already absolute masters of the entire Iraty Forest and the entire slope towards Spain, of our pastures even below this forest. This is how they interpret the unfortunate convention on neutralising the disputed land". Report by M. Salabery to the sub-prefect of Mauléon, op. cit.

[46]General Callier and Baron Gros, letter to the Prefect of the Lower Pyrenees, 22 November 1853; ADPA, coll. 7 M 861.

[47]Treaties of 1862 and 1866, and their additional conventions of 1858, 1863 and 1868.

[48]Ministry of Foreign Affairs, Paris, 21 February 1859; ADPA, coll. 7 M 861.

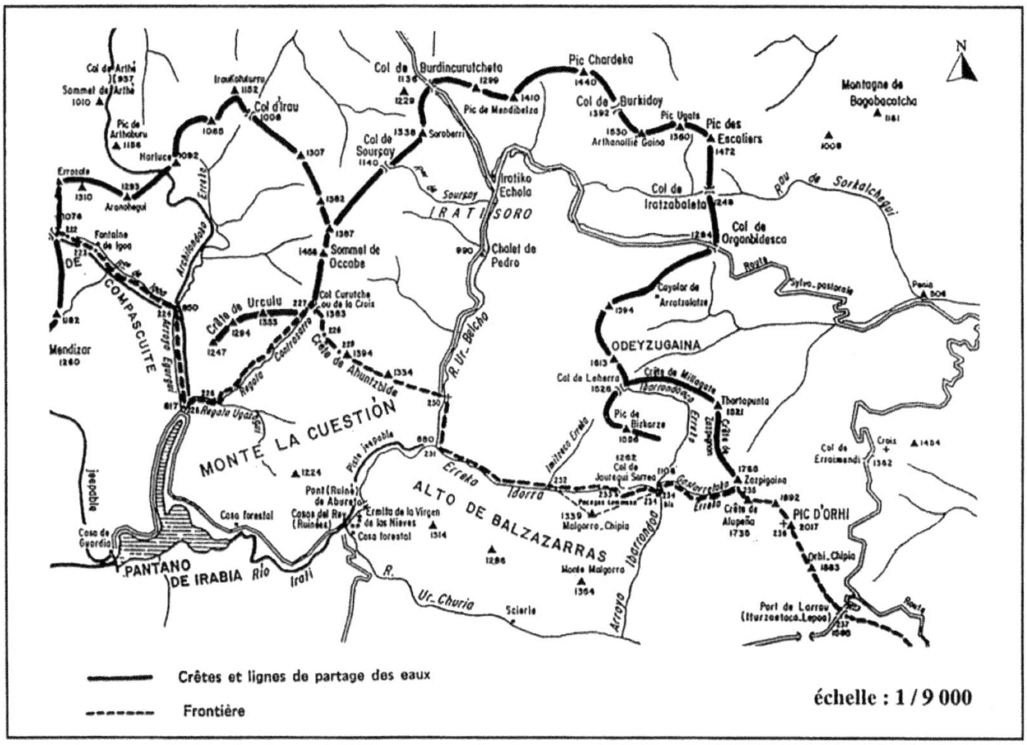

Fig. 7.2 Border demarcation of the Iraty perimeter (Treaty of 1856). *Source* Map by Marie-Thérèse Labarthe (Puyo 2002, p. 77)

In fact, the Iraty massif was at the centre of complex bargaining marked by two different rationales. According to Fernando de Arvizu, the Spanish government was concerned about the high demographic disproportion to its disadvantage in this Pyrenean area; the serious events in Iraty, but above all in the neighbouring region of Quint, were the result of strong pastoral pressure being exerted by the French population, whereas the Spanish side was much more thinly populated. Therefore, from the very beginning of the negotiations, Spain adopted a very rigid position by refusing any revision of the Treaty of 1785 on the Quint region, an agreement that was admittedly favourable to its interests (Arvizu 1992).

As for the French negotiators, their mission was to obtain the maximum pastoral advantage for their side (Arvizu 2001; Puyo 2018), even if it meant accepting concessions in the area of forest resources, as Jean Sermet points out: "Unlike the Parliament of Navarre in Pau after 1785, the French commissioners attached little importance to the Iraty sector, whose pastures to the north of the Errokidor valley (current border) were of interest to the Salazar Valley. Why not seek compensation there for what was going to be claimed from Quint? [...] In fact, the disputed 'monte' [forest massif] was itself of low value [...]" (Sermet 1994, p. 279). However, contrary to that author's opinion, the ceded Iraty forest was not without value—far from it, the compensation received by the Pays de Cize syndicate corresponded to a very large sum. Moreover, Lower Navarre was poor in forest stands at the time, with unallocated forests emerging from several centuries of unbridled exploitation, not to mention the forests of the Aldudes valley, ravaged by 150 years of wood-fired steelmaking (Puyo 2002).

In the end, in Iraty as along the whole Pyrenean chain, the border that had now been established was the result of national and/or local arbitration, favouring either natural boundaries or rights of use, depending on the area (de Arvizu 1997; Capdevila 2009). It should be noted that, as early as 1830, on the occasion of the work of the bipartite Gleizes-Bayona commission, the French commissioner, faced with the thorny question of the 1785 Caro-Ornano demarcation line in the Aldudes valley, which had been strongly questioned by the local communities on the northern slope, put forward a proposal based on a novel barter, founded on a recourse to natural boundaries, in this case the ridge lines: the whole of the Aldudes valley (located on the northern slope) would thus be given to the French, in return for the whole of the Iraty massif (on the southern slope) going to the Navarrese. The French State would also commit to granting financial compensation to the inhabitants of the Pays de Cize, allowing them to rent the pastures "[…] *on the same land* [the Iraty pastures] of *which the Spaniards would have acquired ownership*".[49] According to Commissioner Gleizes, this solution also had the advantage of transferring to Spain a territory which was in fact devoid of French population but which was also considered by him to be of low value due in particular to the difficulty of accessing it. "*Thus, the possessions we have on the southern slope can only be exploited to advantage by the Spaniards […] It should even be noted that the surface area of the land transferred to each other would be more or less the same* [estimated at 6172 *ha* for Iraty], *not including, in the cession that must be made to Spain, the commune of Ondarolle which, in fact, has never ceased to be French and to be part of the Pays de Cize*".[50] But the Spanish very firmly refused to go back on the 1785 agreement, a solution that did not meet with much success, either in 1832 or later, during the work of the bipartite commission set up in 1853.

In our opinion, this paper is a good illustration of what Daniel Norman, who was once questioned about the mechanisms at work that led to the setting of a border under the Ancien Régime, had to say: "It is not possible to say that a frontier is drawn: it is the stakeholders who draw it. They turn the border into a set of political places and drivers that are not exclusively wanted from above, as they result from the complex interplay of vertical relationships between the governors and the constituents. The latter are never entirely passive: they accept, discuss, model, shape. Commissioners are sent in to consult them, or to pretend to do so" (Nordman 2008).

Thus, the research of Jacobo García-Álvarez and Paloma Puente-Lozano (2017a; b), focusing on the demarcation of the Portuguese-Spanish border during the second half of the nineteenth century, has clearly demonstrated how much the territorial vision of local communities of their border space played an important role in the outcome of the final arbitrations made by the bipartite delimitation commissions. And this is an observation that was subsequently taken up by the work of the historian Benjamin Duinat, dedicated to this same border in the western Pyrenees: "[…] no border is totally imposed and the populations never really remain passive before the agencies of the State. Between the centre and the periphery, the analysis must necessarily be multiscalar in order to take into account all the levels of power and thus avoid a simplistic, binary logic of an excluded third party: the intersection of micro and macro perspectives thus gives a fundamental place to the stakeholders and the agency in the study of the effects of encompassing structures" (Duinat 2019, p. 90). Any self-respecting geographer/ planner can only agree.

Archival Sources

ACEGCGE: Archivo Cartográfico y de Estudios Geográficos del Centro Geográfico del Ejército. Madrid

ADMAE: Archives diplomatiques du Ministère des Affaires Étrangères. La Courneuve.

[49]Gleizes, *Observations on the minutes of the boundaries commission, dated 15 March*, 20 March 1831, 1 J art. 105-5, ADPA.

[50]Gleizes, *Notes on the project for a new demarcation of the border between the Aldudes and the Pays de Cize, with the line passing through the ridge separating the two sides of the Pyrenees mountain range*, 5 June 1832, 1 J art. 105-5, ADPA.

ADPA: Archives départementales des Pyrénées-Atlantiques. Pau.
BNF: Bibliothèque Nationale de France, Département Cartes et plans. Paris.

Acknowledgements This work has been supported by the Spanish Ministry of Economy, Industry and Competitiveness and the European Regional Development Fund (Research Project CSO2015-65301-P, MINECO/FEDER), as well as by the Spanish State Research Agency (Research Project PID2020-114088GB-I00/AEI/10.13039/501100011033).

References

Ancel J (1938) Géographie des frontières. Gallimard, Paris

Arvizu F (2001) Frontera y fronterizos: el caso de los Alduides (Pirineo de Navarra). Anuario de Historia del Derecho Español 71:9–48

Arvizu F (1983) Problemas de límites y facerías entre los valles navarros y franceses del Pirineo. Cuadernos de Etnología y Etnografía de Navarra XV:41–42

Arvizu F (1992) El conflicto de los Alduides, Pirineo navarro: estudio institucional de los problemas de límites, pastos y facerías según la documentación inédita de los archivos franceses (siglos XVII-XIX). Gobierno de Navarra, Pamplona

Brocas D, Legaz (2004) La forêt d'Iraty entre XVIe et XIXe siècles: une vue déformée par l'enjeu frontalier ou les débuts d'un mythe. In: Corvol-Dessert A (ed) Les forêts d'Occident du Moyen Âge à nous jours. Presses Universitaires du Mirail, Toulouse, pp 145–152

Capdevila J (2009) Historia del deslinde de la frontera hispano- francesa. Del Tratado de los Pirineos (1659) a los Tratados de Bayona (1856–1868). Centro Nacional de Información Geográfica, Madrid

Cavaillés H (1910) Une fédération pyrénéenne sous l'Ancien régime. Les traités de lies et de passeries. Revue historique 105:1–34

de Arvizu F (1997) La solución institucional del conflicto de los Alduides (Pirineo Navarro). Universidad de León, Léon

de Kock M (1817) Histoire abrégée des traités de paix entre les puissance de l'Europe depuis la paix de Westphalie, tome 3. Gide, Paris

Desplat C (1986) Le parlement de Navarre et la définition de la frontière franco-navarraise à l'extrême fin du XVIIIe siècle. In: Association des Amis de la Bibliothèque centrale de Prêt des Hautes-Pyrénées (ed) Lies et Passeries dans les Pyrénées. Tarbes, pp 109–120

Druenne B (1986): Les lies et passeries, spécialement pendant la guerre de succession d'Espagne. In: Association des Amis de la Bibliothèque centrale de Prêt des Hautes-Pyrénées (ed) Lies et Passeries dans les Pyrénées. Tarbes, pp 69–108

Duinat B (2019) Une anomalie frontalière à la loupe: Ondarrola, un hameau entre la France et l'Espagne (v. 1780 – v. 1860). Revue d'histoire du XIXe siècle 58:249–268

Fain (1828) Manuscrit de l'An Trois (1794–1795) contenant les premières transactions des puissances de l'Europe avec la République française. Imprimerie de Fain, Paris

Foucher M (1986) L'invention des frontières: un modèle géopolitique français. Hérodote 40:54–88

García-Álvarez J (2019) Introducción – La frontera hispano-portuguesa - Representaciones geográficas, paisajes y delimitación territorial (SS. XVII-XXI). Revista de historiografía 23:10–16

García-Álvarez J, Puente-Lozano P (2017a) Bridging central state and local communities' territorial visions: boundary commissions and the making of Iberian borders, 1750–1900. J Hist Geogr 57:52–61

García-Álvarez J, Puente-Lozano P (2017b) A controversial natural border. The making of the Spanish-Portuguese boundary along the Minho River (1855–1866). Geoforum 79:5–16

García-Álvarez J, Puyo J-Y (2019) Los primeros intentos de delimitación moderna de la frontera pirenaica: la comisión franco-española Caro-Ornano y su legado cartográfico (1784–1793). Revista de historiografía 30:15–44. https://doi.org/10.20318/revhisto.2019.4742

García-Álvarez J, Puyo J-Y (2015) La aportación geográfica y cartográfica de las Comisiones de Límites Luso-franco-españolas (siglos XVIII y XIX). Apuntes para una historia comparada del conocimiento y la cooperación científica sobre la frontera. Terra Brasilis (Nova Série), 6. http://terrabrasilis.revues.org/1693

Guichonnet P, Raffestin C (1974) Géographie des frontières. PUF, Paris

Mavidal M, Laurent E (1898) Archives parlementaires de 1787 à 1860, première série (1787–1799), tome 3. Dupont, Paris

Nordman D (1998) Frontières de la France. Gallimard, Paris

Nordman D (2008) La frontière: théories et logiques territoriales en France (XVIe-XVIIIe siècles) - intervention au séminaire Frontafrique du 13 juin 2008. http://www.frontafrique.org/spip.php?article25

Puyo J-Y (2018) De la Bérézina aux Aldudes: le commandant Gleizes et la question épineuse de la délimitation frontalière des Pyrénées occidentales (1827–1832). Stratégique 119:35–62

Puyo J-Y, García-Álvarez J (2016) Une frontière pour les Pyrénées: l'épisode trop méconnu de la commission topographique franco-espagnole Caro-Ornano. Cartes & Géomatique 228:47–58

Puyo J-Y (2002) Faceries et rectifications frontalières: le cas de la forêt d'Iraty. In: Desplat C (ed) Frontières. CTHS, Paris, pp 65–78
Reclus O (1872) Géographie. Paris L. Mulo
Sahlins P (1990) Natural frontiers revisited: France's boundaries since the seventeenth century. Am Hist Rev 95:1423–1451
Sermet J (1994) Le statut original du Pays Quint. Bull pyrénéen 183–184:263–294
Sorel A (1891) L'Europe et la révolution française. Plon, Paris
Zeller G (1933) La Monarchie d'ancien régime et les frontières naturelles. Revue d'histoire moderne 9:305–333

A Water Borderline Adrift Between France and Spain: The Bidasoa and the Bay of Higuer at the End of the Nineteenth Century

8

Benjamin Duinat

Abstract

In February 1893, a violent storm caused an important change at the mouth of the Bidasoa, the contiguous river between France and Spain, so that a rectification of the borderline became an absolute necessity. This 1890s episode reveals all the paradox of a geopolitical limit set according to natural criteria in an aquatic environment marked by continuous movement. It is precisely a *modus vivendi* based on the annual revision of the dividing line in the Bidasoa and the bay of Higuer that made it possible to find a solution to the progressive drift. Delimitation and demarcation of such a moving border area allow to consider the borderline as an institution afflicted with antinomy. In theory at least, a border should remain stable, whereas everything around it is subject to change, customs, economic and social life, the balance of power between states, political regimes, the environment. This episode in the chronicle of border areas shows two things. On the one hand, each level of technology has its own political form and entails certain uses of the border. On the other, establishing a dividing line is not a *spatial given* with consequences for populations who live on either side, but rather an *anthropic construction* from which spatial impact results.

B. Duinat (✉)
University of Pau and Pays de L'Adour, Pau, France
e-mail: benjamin.francois-duinat@univ-pau.fr

8.1 Introduction: A Border Between Fixity and Movement

The intuition was premonitory. In 1858, the writer and journalist Antonio García Gutiérrez gave a detailed analysis in the newspaper *La España* of some of the provisions of the Treaty of Bayonne (1856), which established the western border between Spain and France.[1] His harshest criticisms focused on Article 9, which concerned the Bidasoa, the river separating the two countries at the western end of the boundary. "From Chapitelacoarria [Xapitelako harria] to the mouth of the Bidassoa [Bidasoa], in the Figuier [Higuer] roadstead, the mid-point of the main channel of this river at low tide shall constitute the line separating the two territories".[2] What exactly were García Gutiérrez's objections to it?

[1]For information about Antonio García Gutiérrez (1813–1884), see online the *Diccionario Biográfico Español* (http://dbe.rah.es).

[2]The original French versions of these texts can be found at: ADMAE—308QONT/2; or online: https://basedoc.diplomatie.gouv.fr/exl-php/util/documents/accede_document.php?1600357475257. See also the Spanish version: MAEC (p. 803–812). This extract taken from the English translation in the UN online repository of treaties. https://treaties.un.org/doc/Publication/UNTS/Volume%201142/volume-1142-II-838-English.pdf.

J. García-Álvarez and P. Puente-Lozano (eds.), *Beneath the Lines*, Historical Geography and Geosciences, https://doi.org/10.1007/978-3-030-96904-2_8

> We also disregard the circumstance, which the makers of the treaty consider very important, that the course of the Vidasoa [*sic*], like that of any river not channelled between rocks, is variable. Thus it can, with the inclination of its current, cause the territory or jurisdiction of Spain or France to advance or retreat; although it is assumed this cannot cause serious inconvenience, it will always appear as an informality and lack of fixity in that which should be perennial and permanent, such as a border.[3]

However, in February 1893, a violent storm caused an important change at the mouth of the Bidasoa, making a rectification of the borderline an absolute necessity so that users of the river could freely navigate its waters without the risk of the authorities of the other country boarding their vessels. Such freedom of navigation is, in essence, what the president of the French delegation of the Pyrenees Joint Commission had proposed to the Minister of Foreign Affairs.[4] In addition, the waters at the mouth of the river were full of fish and were therefore coveted.

The supposedly natural dimension of certain sections of a border actually conceals a human artificiality, inasmuch as physical characteristics are only the substratum of mental, ideal and psychological limits (Geouffre de La Pradelle 1928 p. 172). That is why the most important spatial ruptures are in the mind (Gottmann 1952, p. 220). Moreover, so-called natural borders are not exactly barriers, but rather a path, a space for contact and circulation, be it mountain ridges, a river or even the sea (Van Gennep 1995 [1921] p. 161). In other words, the obsession with a clear link between topography and boundaries often leads to the reification, or even to the fetishization, of border space (Fall 2010).

This episode and its aftermath around the mouth of the Bidasoa during the 1890s reveal all the paradox of a geopolitical limit set according to natural criteria in an aquatic environment marked by continuous movement (Donaldson 2011). It is precisely this *modus vivendi*, based on an annual revision of the dividing line in the Bidasoa and the bay of Higuer, which made it possible to find a solution to the progressive drift. It therefore offers us the opportunity to consider the delimitation and demarcation of a moving border area, insofar as the borderline is an institution afflicted by antinomy. In theory at least, it should remain stable, whereas everything around it is subject to change, customs, economic and social life, the balance of power between states, political regimes, the environment and so on (Balibar 2001, p. 174). This paradox and antinomy guide our case study.

From a methodological point of view, ultimately, the analysis feeds on *microhistory* and spatial history, which have recently received two cross-criticisms. On the one hand, space as such is very rarely the object of historical microanalysis. On the other hand, spatial history is generally of a symbolic nature and rarely deals with concrete aspects of the relationship of societies to a space (Torre 2018, pp. 40–46). A study of various sources at the archives of the *Ministère des Affaires Étrangères* (La Courneuve, France) and *Archivo Histórico Nacional* (Madrid, Spain) will provide the opportunity for various levels of observation and make it possible to refute these legitimate accusations. Indeed, many recent works have shown the extent to which the idea of a pure imposition of the border from the top of political society is a misconception; to such an extent that adjustments and accommodations to local realities are a necessary condition for the success of the work of fixing a dividing line (Di Fiore 2017).

First, an analysis of the different devices for fixing the dividing line in international rivers provides the perspective for this study. Secondly, we examine control of the border area with practices of surveying, describing and mapping the space. Finally, the negotiations that led to the *modus vivendi* in 1894 show to the full that in border matters "the devil is in the details": as state agencies refine the scale of action on borders, only then do new problems emerge.

---

[3] *Archivo Histórico Nacional* [AHN]—Ministerio_Exteriores_TR_98. Extract translated from Spanish by the authors.

[4] *Archives du Ministère des Affaires étrangères* [ADMAE]—308QONT/7.

8.2 Fluvial Mischief in Bidasoa

At its source, the Bidasoa is just a stream that originates in Upper Navarre (Spain): the river, at that point called the Baztan, gives its name to a well-known valley community before gradually growing in size with the contribution of several tributaries. From the village of Oronoz to its mouth in the Atlantic Ocean, the river is called Bidasoa. The *Frontière des Pyrénées* map provides a good understanding of the area under analysis (Fig. 8.1). This map shows the entire French-Spanish border on a small scale. It features three areas on a larger scale: the Bidasoa, the Aldudes valley and the enclave of llivia in La Cerdanya. Unfortunately, its author and exact date of creation are unknown, but it certainly dates from after the last international agreement mentioned in the document, 1879.

The water flows towards the south-west, imperfectly drawing an arc of a circle, and then continues its course towards the north-west as far as the ocean. From downstream to upstream: between Cape Higuer in Hondarribia (Fuenterrabía) and the place called Xapitelako harria (Chapitelacoarria), further south, the international boundary follows the river, before breaking off and sketching a route, with a marked sinuosity, through the Pyrenees to the east. In fact, only the last 14 kms of the river, approximately, coincide with the border. It is therefore a contiguous international river, because it forms the border between two sovereign states, and not a successive one, which would first pass through one country and then flow into another. The Rhine, for example, is an alternately successive and contiguous river.

The valley through which the Bidasoa river flows, from the time it enters Guipuzkoa until it reaches Behobia (a little before Hendaye), is very narrow and closed. From this point, as far as the river mouth, the river valley widens and forms a

Fig. 8.1 *Frontière des Pyrénées*, after 1879. *Source* ADMAE—274QO/20. Real dimensions of the document: 59–89 cm

fertile plain, which gradually expands following the direction of the northward flow of water. The river is navigable, for small boats, until near Xapitelako harria, where the Bidasoa ceases to be the dividing line between Spain and France. This is certainly due to the fact that at high tide its waters reach this place (Múgica Zufiria 1918, pp. 23–25).

Between Xapitelako harria and the Atlantic Ocean, the course of the Bidasoa River was the object of a particularly longstanding conflict between the inhabitants of the two riverbanks, as well as between the two monarchies. The history of the dispute dates back to at least the Middle Ages. Successive representatives of the Spanish monarchy maintained that the entire river belonged to them, on the basis that the river had its source in and flowed mainly through Spain. On the other hand, the French defended the idea of shared possession, arguing that solemn acts of Franco-Spanish diplomacy had always taken place in the middle of the Bidasoa River (Descheemaeker 1948).

The chronology of the conflicts and negotiations over the use and ownership of the river revolves around five villages, two in Spain: Irun and Hondarribia, and three in France: Biriatou, Urrugne and Hendaye. Until the middle of the fifteenth century, Hendaye would have been a simple hamlet, which, like the rest of Aquitaine, had belonged to England since the twelfth century. The capture of Bayonne, in 1451, and the end of the Hundred Years' War delivered territories to the King of France, to the detriment of his English rival, that had eluded him for three centuries. Several changes resulted from this: the royal power sought to occupy the territory as effectively as possible to its furthest reaches and encouraged settlement. Thus, the population living in this border area of the Kingdom of France increased, which mechanically exacerbated pressure on the Bidasoa and the resources that it could be expected to provide. Serapio Múgica argues that it was from this period that France and the people living there aspired to possess half of Bidasoa (Múgica Zufiria 1918, pp. 27–28); but perhaps this is simply a question of the information source since the oldest preserved documents attest to the contrary claims of both parties (Legrand 1905, p. 3). Whatever the case may be, the inventory of the disputes between the inhabitants of Hondarribia and Hendaye from 1509 to 1740 occupies 28 full pages in the collection of boundaries kept at La Courneuve.[5]

Curiously enough, Bidasoa escaped the first fever of border fixing at the end of eighteenth century (Nordman 1998; Stopani 2008), but not the second one, dating from the middle of the nineteenth century (García-Álvarez and Puyo 2015). The case of ownership of the river was settled under the first Treaty of Bayonne, in 1856, and that of the bay of Higuer in 1879. According to the first text, the median line marked the division of the waters of the Bidasoa River.

The four possibilities for fixing the borderline in a contiguous international river are as follows: the river forms a kind of no-man's land, and the sovereignty of each country stops on the riverbanks; the entire watercourse belongs to one of the two parties; the thalweg, i.e. the deepest point of the channel, determines the dividing line; finally, the median line, equidistant from both edges, serves as an interstate caesura. The thalweg was chosen in the case of the Rhine or the Vistula, in 1814–1815, but the middle line was chosen in the Bidasoa and the Miño, in the second half of the nineteenth century (Caflisch 1989; Schröter 1992; García-Álvarez and Puente-Lozano 2016). There is no rule in international law. In accordance with the second text, the bay of Higuer is divided into three zones, as the *Frontière des Pyrénées* map shows: one exclusively Spanish, to the west, the other French, to the east, and, finally, a so-called neutral or common zone to allow boats to enter and leave the river without difficulty.[6] "All this neutral space covered at high sea constitutes a kind of border in surface area, replacing the linear

[5]ADMAE—274QO/21, f. 57–71.

[6]The French versions of these texts can be found at: ADMAE—308QONT/2. See also the Spanish versions of these texts (MAEC, vol. 2, t. 2, p. 621–631 and p. 803–812).

border", wrote the French Ministry of the Navy in 1891 about the third zone.[7]

The years 1856 and 1879 mark two decisive moments of this movement of *delimitation*, drawing a line on a cartographic document, *demarcation*, placing markers on the ground, and *maintenance* of the work thus carried out, administering the border area through joint commissions (Jones 1945, p. 57). Nevertheless, these three stages often overlap (Donaldson 2008). However, several problems related to free navigation in the waters of Bidasoa led to the creation of the Joint Commission of the Pyrenees in 1875 (Fernández de Casadevante Romani 1985). It is curious that this institution received that name, although it originally concerned a river area. In any case, the authorities of both countries realised that although the first Treaty of Bayonne (1856) had made it possible to define the dividing line on the Bidasoa, the bay of Higuer remained outside the scope of international agreements. The convention determining ownership of the waters of the bay of Higuer was the fruit of the Commission's work, subsequently validated by the two governments.

This commission was by no means an arbitral institution or an international administrative tribunal, but rather an authority responsible for carrying out investigation into conflicts and disputes between the people living along the Bidasoa River, as well as for application of the 1856 treaty by the two governments and their agents. In 1887, its powers were finally extended to the entire French-Spanish border area.[8] Actually, no sooner had the ink on the final act of the Bayonne Treaties (1868) dried than sharp tensions arose at the mouth of the Bidasoa River.

Number 91 of the *Annonce hydrographique* (1890) fully reproduced the text of the Convention on the French and Spanish jurisdictions, the term here being synonymous with the word territories. A note specified how the joint commission had conducted demarcation of the aquatic area. In accordance with the stipulations of Article 10, the Joint Commission of the Pyrenees marked out and established the lines separating the waters with a series of pyramids or monoliths erected on each side of the bay (Fig. 8.2). The French zone, to the east, lay between the coast of Spain and a broken line given by two alignments. One of these was provided by a rock, R, located on the seaside, south of Cape Higuer, painted black and topped with an iron landmark with a disc painted white, and by the geodetic monument or pyramid marked R', 4 m high, painted black on the seaside and erected on the highest point of the cliff. The other alignment was given by two pyramids marked F' and F", 4 and 6 m high respectively, painted white on the seaside and erected on a cliff north of the village of La Magdalena (Spain). The part of the bay located north of the first alignment, R R', and the part located west of the second alignment, F' F", until its intersection with R R' therefore formed the Spanish area. Finally, the part of the bay between the two French and Spanish zones formed the common or neutral area.[9]

To the extent that changes were possible, the intention of Article 4 of the 1879 Convention was to provide a framework for any changes that Antonio Gutiérrez had feared in 1858:

> Art. 4 – It is understood that the anchorage and the entrance to the river will remain outside the waters under the exclusive jurisdiction of either country. In the event of a change in the position of the sandbar, the following provisions, which fix the distribution of waters as they stand at present, should be amended accordingly.[10]

At the end of 1891, Joaquín Escoriaza, the commander of the ship *Tajo*, stationary in Hondarribia, drew a map of the area for the attention of the Minister of State (Foreign Affairs). This clearly defined in words and cartographic images the issues involved in the coastal drift at the mouth of the Bidasoa River. Already in 1891, the drift of the dividing line was a serious cause for concern. Thanks to this source, it is easy to locate all of the points mentioned in the *Annonce hydrographique* (Fig. 8.3). This *Plano de la*

[7]ADMAE—308QONT/10.

[8]ADMAE—308QO/15. See also MAEC (p. 813–817).

[9]ADMAE—308QONT—10.

[10]ADMAE—308QONT/2. Extract translated from French by the author.

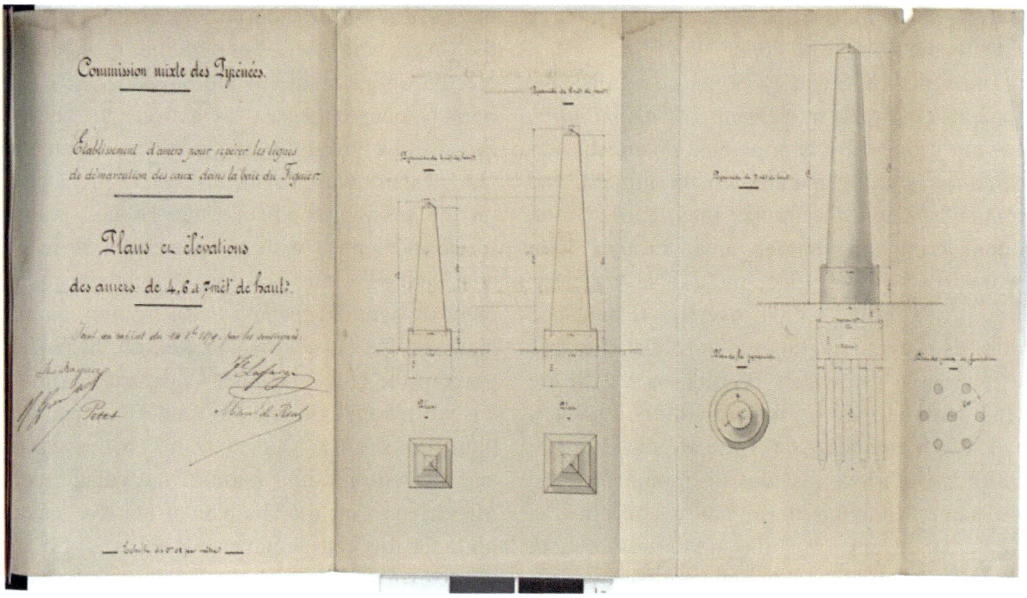

Fig. 8.2 *Plans et élévations des amers*, 1879. *Source* ADMAE—308QO/13. Real dimensions of the document: 35–62 cm

Bahia de Higuer y del último trozo del Bidasoa hasta Endarlaza, changes the conventional orientation, showing the north on the right. The zonal tripartition of the bay at Higuer is clearly visible due to the differentiated representation of the three surfaces. At the bottom of the document, a very large-scale sketch (1/1000) of another part of the river, which is not relevant here, completes this cartographic work.

Because of the displacement of the sandbar and the channel at the mouth of the Bidasoa, this military man expressed the desire to revise the dividing lines in the bay of Higuer. The main disadvantage noted by Joaquín Escoriaza was that neutral waters bathed the Spanish coast. The rectification he requested, but did not obtain at the time, was to grant Spain an exclusive strip of water contiguous to its coastline.[11] It was ultimately the violent storm of February 1893 that made changing the delimitation lines of the waters in the bay of Higuer an absolute necessity.

8.3 Surveying, Describing and Mapping an Aquatic Space

In March 1893, the president of the French delegation of the Joint Commission of the Pyrenees sent a complete dossier to the Minister of Foreign Affairs. In it, he mentioned the storm that had led to a change in the river mouth and the fact that the commanders of the stationary ships were studying the issue. Both commanders travelled through the area by boat, as they used to on a regular basis. For example, Joaquín Escoriaza did so in 1891, when he wrote a complete report, translated into French, on the state of the Bidasoa, its tides, navigation, fishing and, of course, the perpetual movement of the sandbar and the channel. As in the negotiations leading up to the 1879 agreement, the two commanders used a French map of the harbour drawn up in 1866–1867. This exemplifies precisely one of the aspects of a historiographic trend that focuses on the study of the "behind-the-scenes" of maps, which pays less attention to an analysis of their content and more to the sum of operations and the consultative practices employed in their production (Laboulais 2008). This perspective

[11] AHN—Ministerio_Exteriores_TR_194.

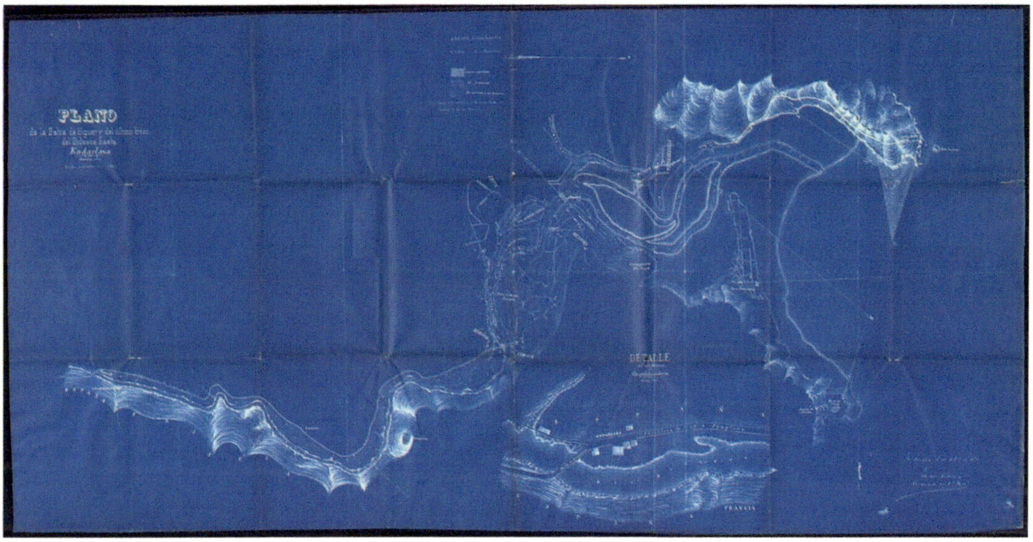

Fig. 8.3 *Plano de la Bahia de Higuer y del último trozo del Bidasoa hasta Endarlaza*, 1891 (scale: 1/10,000). *Source* AHN—Ministerio_Exteriores_MPD_35. Real dimensions of the document: 68–135 cm

renews attention to surveys and information gathering in the field, to moments of pragmatic adaptation and adjustment, as well as to the malleability of using the map according to needs. Thus, it avoids reifying the omnipotence of maps over political practices, without undermining the importance and specificity of these documents in terms of government and, therefore, the affirmation of an authority over the territory (Blais 2014, pp. 10–11).

Maps are thus envisaged as a means of providing evidence, particularly in cases of dispute, and as a support for negotiation intended to reinforce the point of view defended. Maps circulate among and receive annotations and comments from the various levels of power that form a group of educated perspectives on the graphic grammar of space. A map acquires meaning only within a scholarly or technical community that has produced it, within a social community that consults it or exhibits it, that uses it for the rhetoric of power that it deploys (Lepetit 1996). However, in these matters, border populations are not ignorant or incompetent and do not remain on the fringes of cartographic use. Numerous examples show that during surveys carried out by the Joint Commission, border inhabitants were perfectly capable of reading a map, indicating places on it, or simulating movements in the space with gesture.[12] This example allows us to return to the criticisms sometimes levelled at maps in the context of border negotiations. The abstraction, use of mysterious figurative language or the staging of ideal lines that are never found in the field, to which must be added the confinement of these documents in state archives, limiting consultation to a very closed circle of the elite, would result in cartographic documents that lack the persuasive capacity sometimes attributed to them. This is, in essence, what an Italian geographer had to say about the maps used during the negotiations of the Treaty of Utrecht in 1713 (Sereno 1999). However, the uses of maps developed considerably during the eighteenth and especially nineteenth centuries, so the circumstances were quite different by the 1890s (Verdier 2015).

The example of the work carried out in parallel by the two commanders of the Spanish and French navies offers the opportunity for close analysis of this behind-the-scenes approach to the map-making process. Thus, in March 1893,

[12]ADMAE—308QO/13.

Julien Viaud, *alias* the writer Pierre Loti, sent a text and a map that he had reused to the president of the French delegation of the Joint Commission, who in turn sent it to the Minister of Foreign Affairs. State knowledge gathered at ground level, in the field, circulated from the choppy waves of a bay on the Atlantic coast to ministerial cabinets. The basis for Viaud's description was a map produced by a French *École Polytechnique*-educated hydrographic engineer, Anatole Bouquet de la Grye, in 1866–1867. In order to highlight the results of the research carried out by each party, Julien Viaud took the liberty of colouring the small space that was the object of so much attention on the map by means of surface figures (Fig. 8.4). His observations are the result of an attentive examination of this delicate space:

> The blue-tinted part on the map represents the French waters in the bay of Higuer; the yellow-tinted part, the Spanish waters; the pink-tinted part, the area that Spain is currently asking us for.
>
> Lines AB and A'B' represent the current direction to enter or leave the river. The blue line AB is the result of my personal observations; line A'B' is the result of Commander [...] Escoriaza's observations. The small and insignificant difference between the two is that we did not use the same alignments, perhaps also because we conducted our observations one week apart. The two directions are equally good to follow; only, one should not deviate either to the right or to the left from the interval they form.
>
> The double blue line A"B" represents the direction that had to be followed last October [1892], when Mr. Escoriaza and I observed it together. And finally, the black line marked A'"B'" represents approximately the direction that had to be followed in the past, along the Spanish coast, at the time when the waters were divided [in 1879].
>
> The entrance passage, after having taken about thirty years to move from A'"B'" to A"B", has therefore suddenly retrogressed this winter to A'B' and AB [...].[13]

Commander Viaud concluded that the Spanish delegation's request on this point was perfectly justified and admissible, as the extent of the neutral waters would guarantee free navigation. After having read the report and seen Commander Viaud's reworked map, the French Minister of the Navy prudently suggested that it would be necessary to wait to see how the river's thalweg, whose channel seemed to be returning to its ancient location, evolved before agreeing on a new delimitation that could quickly prove to be unsuited to its environment.[14] Viaud based his opinion on the absence of obstacles to free navigation. However, a few months later, after carefully observing the behaviour of the shoals of fish, he changed his mind, as the salmon in particular entered and left the Bidasoa only through the stretch of water demanded by Spain, namely the old passage. However, despite salmon fishing not being a particularly lucrative activity in terms of quantity, the fact remained that local inhabitants had a strong connection to it. Viaud was therefore well aware that a challenge to equality of access to this resource, even if symbolic, would lead to heated disputes, which it was the primary function of the Joint Commission to curb, control and eliminate.[15] For the time being, the French side therefore decided to temporize.

In any case, these discussions between the field and all levels of power show the extent to which moving through a space, its precise description and the use of cartographic documents are part of the practices and mechanisms of government associated with the consolidation of nation states and administrative rationalisation (Blais 2017). The operations necessary for the demarcation of space are based on "genealogies of places" and "legends of territories" (de Certeau, 1990 [1980], p. 180). State agents go to the places in dispute to hear the different versions, compare title deeds and documents specific to each community of inhabitants, seek conciliation and, finally, decide when the claims overlap. It is thus a question of distributing the space coveted by several groups of individuals in a competitive situation.

[13]Several copies of this document exist: ADMAE—308QONT/10 and 308QO/18. Extract translated from French by the author.

[14]Ibidem.

[15]ADMAE—308QONT/10.

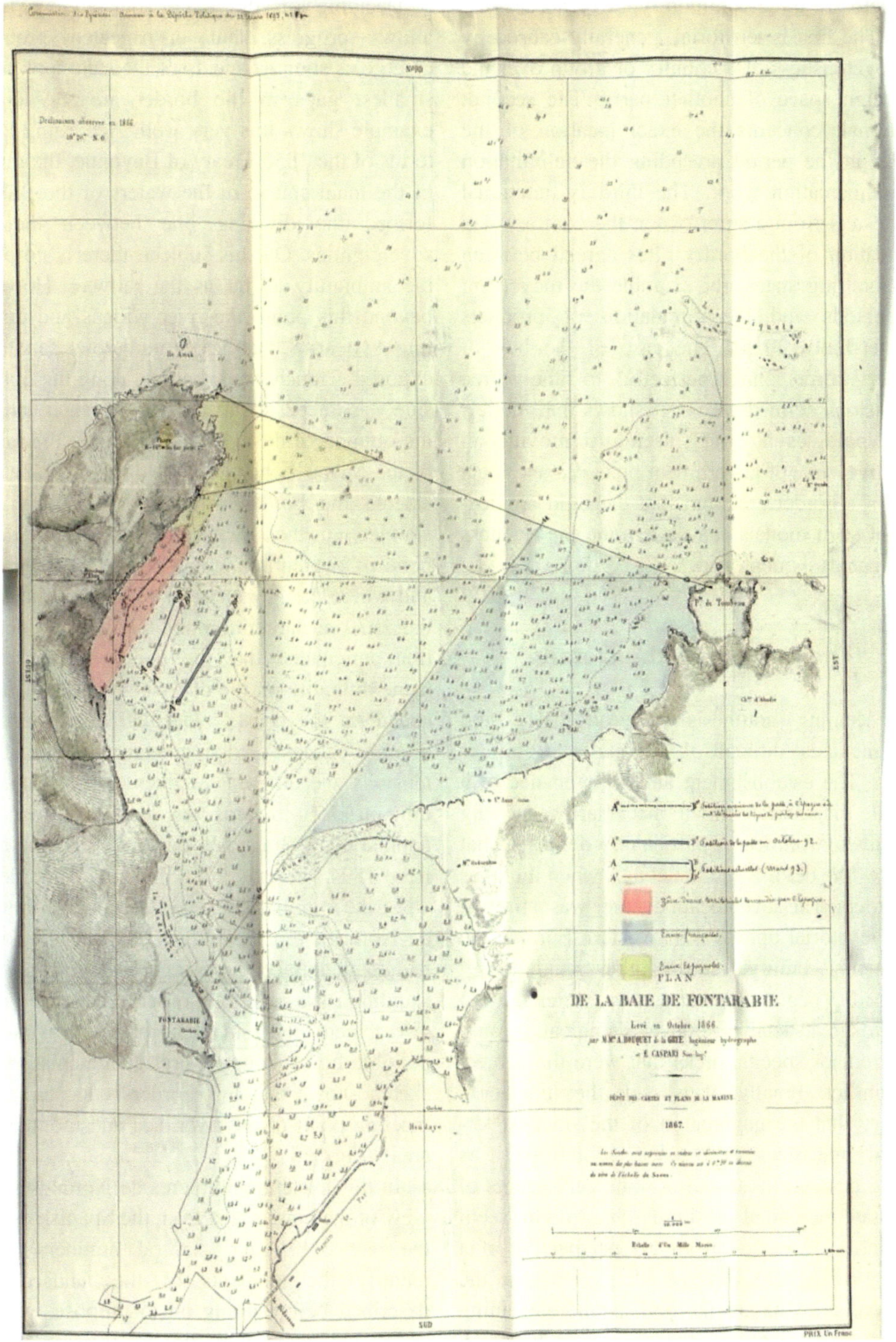

Fig. 8.4 *Plan de la baie de Fontarabie*, 1866–1867 (scale: 1/10 000). *Source* ADMAE—308QO/18. Real dimensions of the document: 52–80 cm

There are three common types of border dispute. The first is territorial, generally caused by the covetousness of a country or group over the sovereign space of another party. The second, positional, concerns the exact location of the border in the period preceding the delimitation and demarcation work. The third is functional and results from a strict or, on the contrary, lax application of the border rules agreed between neighbouring states, for example the movement of persons and the circulation of products (Brunet-Jailly 2015). The case of the bay of Higuer corresponds perfectly to the three dimensions defined above. It shows that models and typologies are very useful to provide an objective view of government practices and local uses of a border area, but that the rigid application of such models can undermine the richness of the analysis deployed.

8.4 "The Devil is in the Details"

The seeming antithesis between fixity and movement is difficult to reconcile in border affairs. The establishment and maintenance of a spatial disjunction in a necessarily moving aquatic environment is particularly delicate. That is why "the devil is in the details" when it comes to borders. The seventeenth century was a time of great territorial upheaval in Western Europe; the eighteenth century, on the other hand, corresponded to negotiations of sectors, which did not go into great detail insofar as demarcation only occurred for specific areas that were the subject of conflict. Finally, it was in the nineteenth century that the negotiation of the smallest territorial interstices took place. Moreover, it is sometimes curious to see how fiercely agents of the state defended positions that might seem laughable. However, from the perspective of a history of border areas, the attention paid to the status of a tiny hamlet has perhaps more meaning than massive territorial displacements (Nordman 1998, pp. 285–286). The power, or, on the contrary, the weakness displayed in border areas is a sign that betrays the real degree of strength of the state (Toubert 1992, p. 16).

Deciding on the major outstanding issues allows progress, but unsuspected problems emerge as soon as agents of the state look at the smallest gaps in the border area. A double example shows this very well. According to the terms of the 1856 Treaty of Bayonne, the middle of the main course of the waters of the Bidasoa formed the dividing line between the two sovereignties. On this subject, there is no possible ambiguity as far as the railway. However, beyond this point, the river widens and divides into two arms that leave numerous sandbanks exposed. The channel running along the Spanish side was deeper than that of the French bank. In the opinion of the French delegation, the middle of the Spanish channel should therefore be considered the border, whereas the Spanish side claimed it as their own territorial waters. Another bone of contention was where the river begins and where the bay ends, to which the clauses of the 1879 declaration apply. The French delegation anticipated that the question might raise some difficulty on the day when work was to go ahead on the coast.[16] Detailed discussions are often a catalyst and a revelation of the latent tensions and different regimes of spatiality. The difficult path towards the *modus vivendi*, first applied in 1893 and subsequently taken up again from 1894 onwards, fully illustrates this idea.

In the summer of 1893, Grégoire Coffinières de Nordeck replaced Julien Viaud. The former had to undertake the task of studying the outstanding issue in situ. In order to help the French government understand the problem as well as possible he drew up several documents. First, a sketch shows that this border is located at the meeting point of the river, sea and mountainous areas (Fig. 8.5).

In this sketch, Coffinières de Nordeck gave a view of the border area from the Spanish coast of the bay of Higuer. Several numbers clearly identify the elements of the landscape he described before going on to propose two possible solutions. Only the main points of this document will be highlighted: the number 5 corresponds to the passage fishermen used on

[16]ADMAE—308QONT/7.

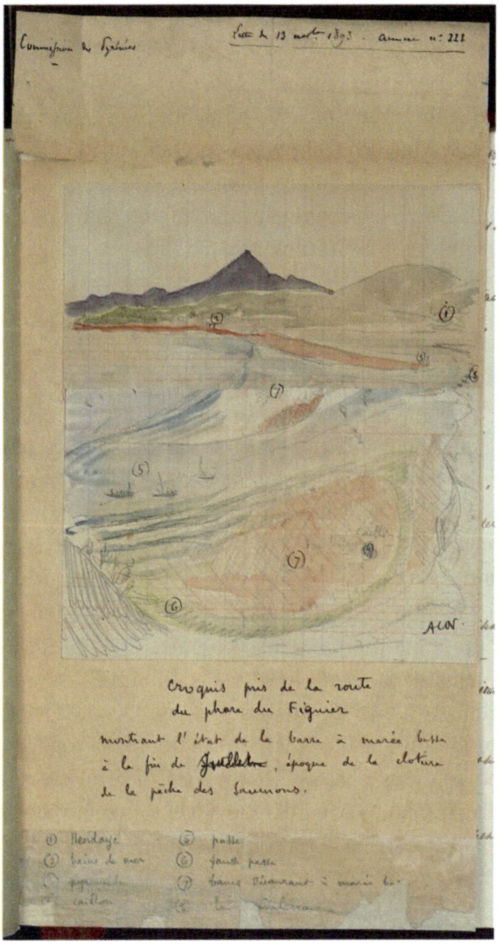

Fig. 8.5 *Croquis pris de la route du phare du Figuier*, 1893. *Source* ADMAE—308QO/18. Real dimensions of the document: 19–36 cm

both sides of the river; on the opposite side, 6 represents the false passage, unusable at low tide because of the sandbanks which were then discovered, designated by 7. Nevertheless, Coffinières de Nordeck made two proposals, one definitive and another consisting of an annual revision of the dividing line at the opening of the fishing season:

> In order to do this, during the month of January, 'seamarks' would have to be built on the Spanish coast to mark this alignment of limits. These seamarks would be permanently built, if my first way of looking at it were adopted, whereas, in the second case, they would be solemnly destroyed to be officially rebuilt before fishing began the following year. To show that these seamarks are only temporary and to distinguish them from those that have been definitively erected, it could be agreed to paint them red, while the others would be painted white.[17].

In both cases, the president of the French delegation of the Joint Commission, Clavery, specified to the Minister of Foreign Affairs that the anchorage and the entrance to the river, in accordance with the letter of the 1879 convention, would remain outside the waters under the exclusive jurisdiction of either country. In his view, the Spanish delegation was not inclined to accept the final solution proposed by Coffinières de Nordeck, as this would amount to agreeing to a new delimitation which was less advantageous than the one suggested by the Spanish Government. Similarly, Spain would not be inclined to wait, as proposed by the French Minister of the Navy, to see how the mouth of the Bidasoa River developed after the abrupt change in 1893. Only Coffinières de Nordeck's second proposal seemed to be acceptable to Spain, according to Clavery: an interim provision, which would be subject to annual review, pending the results of the experiment; this would be "a kind of *modus vivendi*".[18]

As noted in the minutes of the Joint Commission, the Spanish delegation accepted the principle of the annual revision of the boundary between Spanish waters and the neutral zone.[19] How did they proceed? In 1893, as a matter of urgency in order to avoid conflict between the inhabitants of both sides, the commanders of the stationary ships of the French and Spanish navies carried out this delimitation on a cartographic document, and provisionally demarcated the area in question on the ground. This new limit on the water was therefore subordinate to the movement of the bay's depths. The line determined contradictorily by the two commanders remained fixed for a little less than a year, before being

[17]Several copies of this document exist: ADMAE—308QONT/10 and 308QO/18. Extract translated from French by the authors.

[18]ADMAE—308QO/18. Extract translated from French by the author.

[19]Ibidem.

systematically modified according to environmental circumstances.

For several years, the balance was maintained. However, during the winter of 1899–1900, violent weather conditions once again significantly changed the situation at the mouth of the river, so that for a while the *modus vivendi* seemed no longer to satisfy the Spanish side. The president of the Spanish delegation expressed his annoyance to his minister, after the conferences held in Bayonne in 1900 with its French counterparts.

> [...] there is a considerable difference of opinion between the Spanish and French commanders of the stationary ships about the procedure to be used to find the dividing line in accordance with the 1894 "modus vivendi", and if the criterion upheld [...] prevails, it would result in the reduction of our jurisdictional waters with regard to the surveillance and practice of fishing.[20]

The basis for this opinion was a report drawn up by the Spanish commander. In his view, the continuous change of the sandbar due to the winter floods and storms did not provide a solid basis that would support official negotiations, and limited any agreements to friendly, ad hoc, arrangements between the commanders, with proposals of measures whenever the sandbanks shifted.[21]

This summarises the paradox of the moving fixity of an aquatic frontier. This type of arrangement, providing for changing riverbanks, had already been applied between Belgium and the Netherlands in 1843 for the river Meuse. But that case was very different to the one planned for the Bidasoa. The closest precedent is the Treaty of 1827 between France and Baden, which provided that each year in October an ad hoc commission was to ascertain the position of the thalweg, the line of which was to serve as the limit for one year (Schröter 1992, pp. 970–971). In the case of the Bidasoa, it was ultimately the urban development and planning carried out on the river banks and channel at the river mouth at the beginning of the twentieth century that made it possible to put an end to this annual adjustment of a small but decisive part of the French-Spanish border.[22] In the 1959 convention on fishing in the Bidasoa River and the bay of Higuer, the line delimiting the Spanish and neutral zones became fixed. Moreover, there was nothing in the text to provide for any future changes.[23] Management of the natural environment was an absolute prerequisite for the fulfilment of a fixist concept of interstate boundaries.

8.5 Conclusions

The concept of agency i.e. the capacity of the humble to act in constrained situations (Thompson 1957), and hidden transcript as the strength of the weak (Scott 1990) are frequent considerations, whether implicit or explicit, when establishing borders. The work on fixing borders in Europe and in European colonies illustrates that agents of the state have understood the *sine qua non* for success is to take into account both conditions in the field and the customs of local populations (Blais 2014 p. 197–237; Lefebvre 2015; García-Álvarez and Puente-Lozano 2015; Vaillot, 2021). A counter-example serves to show the extent of this idea: when, in 1785, France and Spain decided to establish an absolute borderline in Aldudes, an undivided valley between Upper and Lower Navarre, prohibiting the local practice of shared pastureland the result was a total fiasco. The authorities never succeeded in enforcing its application, or even in maintaining the 1787 demarcation, which had been the object of acts of vandalism. For the Bidasoa, the point of view of agents of the state, whether in the field or in ministerial cabinets, was taken into account. However, they listened to the local population and considered whether the planned measures were likely to receive acceptance.

[20] AHN—Ministerio_Exteriores_TR_194. Extract translated from Spanish by the author.

[21] Ibidem.

[22] ADMAE—296QO/3.

[23] The French version of this text can be consulted online: https://basedoc.diplomatie.gouv.fr/exl-php/util/documents/accede_document.php?1600247117569. To consult the Spanish one (MAEC, vol. 2, t. 2, p. 1163–1184).

The specifics of a "liquid" border upset many of the fixed grids of the analysis of borders and border areas (Morieux 2016). The wealth of this area of study has attracted the attention of researchers who are trying, in particular, to build a database on international rivers to gain perspective on the theories and practices of the division of space in an aquatic environment (Donaldson 2009). The task is arduous and indispensable, but it does not replace an analysis of archival documents, which are the only source that can give substance to the work of geo-historians. The heart of border tensions, the daily banality of the lives of local inhabitants, at ground level, and the transfer of knowledge and skills, as well as the transposition of models to other spaces, from the point of view of commissions and states, can only be seen in the complexity of the primary sources. In this respect, comparative research can be very productive. A case in point is Franco-Spanish negotiations about the Muni, the contiguous river separating Equatorial Guinea from Gabon, in 1900. The Bidasoa served as a reference for the plenipotentiaries of the two countries, even though in the end they settled on the thalweg and not the median line as the limit in the 1900 Treaty of Paris (Mabire 1989).[24]

As is often the case, the apparent heterogeneity of the prose of state-led inquiries is literally obliterated by the unity of the object described, explained, mapped. In the thirst for knowledge, there are many dynamic interactions between states and societies: if we consider the former as an organism set in motion by agents and their practices, nineteenth century European states are partly a product of the culture of inquiry. At the same time, governed societies end up thinking of themselves in terms of the categories employed in the course of inquiries; the information gathered nevertheless remains co-produced by local populations who do their utmost to keep a free hand (Karila-Cohen 2010). This heterogeneous convergence applies perfectly to the issue of delimitation commissions and commissions for the observance of borderline standards (Cairo and Godinho 2013; García-Álvarez and Puente-Lozano 2017). In other words, each stage of the technique corresponds to possible border forms (Carassi, 1996). That is why the tortuous path towards this *modus vivendi* of 1894 shows that a dividing line is not a *spatial given* with consequences for populations who live around it, but rather an *anthropic construction* from which spatial impacts result (Simmel 2013 [1908], p. 607).[25]

[24]AHN—Ministerio_Exteriores_TR_321.

Archival Sources

ADMAE: Archives diplomatiques du Ministère des Affaires Étrangères. La Courneuve.
AHN: Archivo Histórico Nacional. Madrid.

References

Balibar É (2001) Nous, citoyens d'Europe? Les frontières, l'État, le peuple. La Découverte, Paris
Blais H (2014) Mirages de la carte. L'invention de l'Algérie coloniale. Fayard, Paris
Blais H (2017) Carte. Singaravélou P, Venayre S. Histoire du monde au XIXe siècle. Fayard, Paris, pp 434–438
Brunet-Jailly E (ed) (2015) Border disputes. A global Encyclopedia. ABC—CLIO, Santa Barbara (California)
Caflisch L (1989) Règles générales du droit des cours d'eau internationaux. Collected Courses of The Hague Academy of International Law 219, pp 9–225
Cairo H and Gondinho P (2013) El Tratado de Lisboa de 1864: la demarcación de la frontera y las identificaciones nacionales. Historia y Política 30:23–54
Carassi M (1996) Topographes et diplomates: l'art de délimiter les frontières. In: Carassi M, Cusanno C (eds) Massabò Ricci I. Securitas et tranquillitas Europae Ministero per i Beni Culturali e Ambientali, Turin, pp 190–215
De Certeau M (1990 [1980]) L'Invention du quotidien. 1. Arts de faire. Gallimard, Paris
Fernandez de Casadevante Romani C (1985) La frontera hispano-francesa y las relaciones de vecindad (especial referencia al sector fronterizo del País Vasco). Universidad del País Vasco/Euskal Herriko Unibertsitatea, Bilbao

[25]This formula is loosely based on a remark by the sociologist Georg Simmel on the subject of borders.

Descheemaeker J (1948) La Bidassoa et l'Île de la Conférence. Eusko-Jakintza: revue d'études basques – revista de estudios vascos 2(6):649–680

Di Fiore L (2017) The Production of borders in nineteenth-century Europe: between institutional boundaries and transnational practices of space. Eur Hist Rev 24(1):36–57

Donaldson JW (2008) Politics and scale in boundary-making: the work of boundary commissions. J Hist Geogr 34(3):393–396

Donaldson JW (2009) Where rivers and boundaries meet: building the international river boundaries database. Water Policy 11:629–644

Donaldson JW (2011) Paradox of the moving boundary: legal heredity of river accretion and avulsion. Water Altern 4(2):155–170

Fall J (2010) Artificial States? On the enduring geographical myth of natural borders. Polit Geogr 29:140–147

García-Álvarez J and Puente-Lozano P (2015) Las Comisiones Mixtas de Límites y las representaciones geográficas de la frontera hispano-portuguesa (1855–1906). Revista de Historiografía 23:67–100

García-Álvarez J and Puyo J-Y (2015) La aportación geográfica y cartográfica de las Comisiones de Límites luso-franco-españolas (siglos XVIII y XIX). Terra Brasilis (Nova Série) 6, online: https://journals.openedition.org/terrabrasilis/1693

García-Álvarez J, Puente-Lozano P (2016) La frontera hispano-portuguesa en el río Miño en el Tratado de Límites de 1864. Deslinde y ordenación de un territorio fluvial conflictivo. In: Vera Rebollo JF, Olcina Cantos J, Hernández M (eds) Paisaje, cultura territorial y vivencia de la geografía. Libro homenaje al profesor Alfredo Morales Gil. Universitat d'Alacant, Alicante, pp 761–788

García-Álvarez J, Puente-Lozano P (2017) Bridging central state and local communities' territorial visions: boundary commissions and the making of Iberian borders (1750–1900). J Hist Geogr 57:52–61

Geouffre de La Pradelle P (1928) La Frontière. Éditions internationales, Paris

Gottmann J (1952) La Politique des États et leur géographie. Armand Colin, Paris

Jones SB (1945) Boundary-making: a handbook for statesmen, treaty editors and boundary commissioners. Carnegie Endowment for International Peace, Washington, DC

Karila-Cohen P (2010) État et enquête au XIXe siècle: d'une autorité à l'autre. Romantisme 149:25–37

Laboulais I (2008) Introduction. Les "coulisses" des cartes à l'époque moderne. In: Laboulais I (ed) Les Usages des cartes (XVIIe-XIXe siècle). Pour une approche pragmatique des productions cartographiques, PUS, Strasbourg, pp 5–14

Lefebvre C (2015) Frontières de sable, frontières de papier. Histoire de territoires et de frontières, du jihad de Sokoto à la colonisation française du Niger (XIXe-XXe siècles). Éditions de la Sorbonne, Paris

Legrand T (1905) Essai sur les différends de Fontarabie avec le Labourd, du XVe au XVIIIe siècle. Revue du Béarn et du Pays Basque 1–94

Lepetit B (1996) Compte-rendu de Christian Jacob, L'Empire des cartes. Approche théorique de la cartographie à travers l'histoire. Annales. Histoire, Sciences Sociales 4:907–909

Mabire J-C (1989) Elobey Grande, un différend frontalier franco-espagnol. Mélanges de la Casa de Velázquez 25:517–526

MAEC (2006) Acuerdos fronterizos con Portugal y Francia. Colección de recopilaciones, vol. 2, t. 2, 1659–2004, Francia. Ministerio de Asuntos Exteriores y de Cooperación, Madrid

Morieux R (2016) The Channel. England, France and the Construction of a Maritime Border in the Eighteeth Century. Cambridge University Press, Cambridge

Múgica Zufiria S (1918) Guipúzcoa. In: Carreras i Candi F (ed) Geografía General del País Vasco-Navarro, vol 6, Alberto Martín, Barcelona

Nordman D (1998) Frontières de France. De l'espace au territoire (XVIe-XIXe siècle). Gallimard, Paris

Schröter F (1992) La délimitation dans les fleuves internationaux. Annuaire Français De Droit International 38:948–982

Scott JC (1990) Domination and the arts of resistance: hidden transcripts. Yale University Press, New Haven

Sereno P (1999) La costruzione di una frontiera: ordimenti territoriali nelle Alpi Occidentali in età moderna. In: Gregoli F, Simoneta Imarisio C (eds) Le Alpi occidentali da margine a cerniera: le nuove frontiere della geografia, una geografia senza frontiere. Cortina, Turin, pp 75–93

Simmel G (2013 [1908]) Sociologie. Études sur les formes de la socialisation. PUF, Paris

Stopani A (2008) La Production des frontières. État et communautés en Toscane (XVIe-XVIIIe siècles). École française de Rome, Rome

Thompson EP (1957) Socialist humanism: an Epistle to the Philistines. The new reasoner. A Quart J Socialist Humanism 1:105–143

Torre A (2018) Micro/macro: ¿local/global? El problema de la localidad en una historia espacializada. Historia Crítica 69:37–67

Toubert P (1992) Frontière et frontières: un objet historique. In: Posson JM (ed) Castrum 4. Frontière et peuplement dans le monde méditerranéen au Moyen âge. École française de Rome – Casa de Velázquez, Rome-Madrid, pp 9–17

Vaillot B (2021) Aux portes de la nation. Une histoire par en bas de la frontière franco-allemande (1871–1914). PhD, European University Institute, Florence

Van Gennep A (1995) [1921]) Traité comparatif des nationalités. CTHS, Paris

Verdier N (2015) Entre diffusion de la carte et affirmation des savoirs géographiques en France. Les paradoxes de la mise en place de la carte géographique au XVIIIe siècle. L'Espace géographique 44(1):38–56

The Demarcation of the Border of the Spanish Protectorate in Morocco (1913–1931): Chronicle of a Failure

9

Luis Urteaga

Abstract

During the 1910s and 1920s, an international boundary commission made three successive attempts to delimit the French–Spanish border in Morocco on the ground. The first of these, conducted from 1913 to 1914, was partially successful in that it succeeded in demarcating approximately one third of the border. The next two attempts, respectively conducted between 1919 and 1921, and between 1926 and 1931, ended in resounding failure. Based on unpublished cartographic and documentary material, this study examines the field work carried out on Moroccan territory by the Spanish section of the Spanish–French Boundary Commission, assessing the challenges involved in carrying out the cartographic work and why it was discontinued, and discussing the reasons its efforts were thwarted. In Spain, awareness of the failure of these delimitation efforts has fueled an extensive historiographical literature, whose common theme is to deplore France for its diplomatic intransigence and its voracious colonial expansionism. The cartographic sources used to support this thesis are identified and analyzed.

L. Urteaga (✉)
University of Barcelona, Barcelona, Spain
e-mail: urteagago@gmail.com

9.1 Introduction

Problems relating to the delimitation and demarcation of borders featured prominently in the colonial agenda during the final stages of the scramble for Africa, especially in the early decades of the twentieth century. Imperial powers mustered all of their diplomatic might to negotiate and agree on the establishment of new colonies. Nevertheless, territorial allocation plans, which were drawn up after arduous negotiations in European foreign ministries, were not always attune to the reality that was to be found in Africa (Butlin 2009). Colonial agents found that borders were a hotbed of dispute and conflict. The seemingly unequivocal lines drawn up in treaties became uncertain boundaries in the landscape.

The border between the French and Spanish protectorates in Morocco offers a good example of the tensions inherent in the process of colonial partition, and of the role of cartography in attempts to resolve delimitation issues. Between 1913 and 1931, as many as three international commissions sought to delimit the French–Spanish border in Morocco on the ground. The first of these, which was active from 1913 to 1914, was partially successful, in that it succeeded in demarcating approximately one third of the border. The following two, which operated between 1919 and 1921, and between 1926 and 1931, respectively, ended in unmitigated failure.

J. García-Álvarez and P. Puente-Lozano (eds.), *Beneath the Lines*, Historical Geography and Geosciences, https://doi.org/10.1007/978-3-030-96904-2_9

In Spain, awareness of this failure has fueled an extensive academic literature, whose common theme is to deplore the intransigence of French diplomacy and to denounce our neighbors' imperial voracity. Paradoxically, however, very little effort has been devoted to studying how exactly attempts were made to determine the border in Moroccan territory and, in particular, to the importance of boundary commissions in this process.

This research, which essentially relies on unpublished cartographic and documentary material,[1] looks at the field work conducted in Morocco by the French–Spanish boundary commissions. The first section describes the usual sequence of operations in the demarcation of colonial borders and considers the territorial allocation that ensued from the Spanish–French agreement of 1912. The second section presents the structure and duties of the Boundary Commission of Morocco. The third, fourth, and fifth sections successively address the delimitation attempts carried out in the Loukkos region, in the Guerrouaou plateau, and in the foothills of the Rif. The chapter closes with some conclusions and the list of bibliographical references.

9.2 Colonial Borders

Border delimitation operations were among the most common practices of colonial cartography, but they were also among the most problematic. The demarcation of borders amounted primarily to a practical prerequisite for proper administration, particularly in recently colonized areas. But, it was also a political necessity. Borders amounted to a sign of sovereignty, a public assertion of control over a territory. In a sense, the mere presence of cartographers on the ground had a performative and legitimizing value: It made the metropole present in the landscape of the colony. The agreed delimitation of boundaries was the ideal means to minimize potential conflicts between bordering states (Donaldson 2008a).

On the eve of the First World War, the modern practice of establishing borders, partly shaped through the colonial experience itself, was firmly established. According to one key author on the subject, the process may be divided into four stages (Jones 1943). The first stage, essentially political in nature, involves territorial allocation, or the partition of territories by means of a general agreement on where the line should run, enshrined in an international agreement or treaty. In this initial phase, political consensus prevails over precision in determining the geographical location of the border. The second stage, which of delimitation, involves a description of the border in sufficient detail for it to be mapped on a large scale. During this stage, which is technical in nature and usually carried out on the ground, an attempt is made to reconcile inadequate textual definitions in treaties with the local geographical reality. The third stage, which of demarcation, involves the physical materialization of the course of the border using boundary markers or signage. Finally, once the border has been demarcated, it must be maintained through the adequate maintenance of signage and the possible readjustment of boundaries.

In theory, each of the above stages forms the basis of the next one. However, in practice, the process of defining borders did not always follow such a linear path. It was possible for different phases to overlap, with delimitation and demarcation operations running in parallel, for instance. Delimitation work, in turn, could be used to improve knowledge of the terrain, which had been weak up until then, thus prompting a readjustment of the territorial allocation after a fresh round of diplomatic negotiations.

Empirical studies highlight marked differences between the intra-European experience and that

[1]In preparing this study, we used sources from the Archivo General de la Administración de Alcalá de Henares (AGAH), the Archivo General Militar de Madrid (AGMM), the Archivo General Militar de Segovia (AGMS), the Archivo Cartográfico y de Estudios Geográficos del Centro Geográfico del Ejército (ACEGCGE), the Biblioteca del Instituto Geográfico Nacional (IGNM), the Archivo del Ministerio de Asuntos Exteriores y de Cooperación (AMAEC), the private archive of the Jiménez Ortoneda family (APJO), and the Arxiu Històric de la Cambra de Comerç de Barcelona (AHCCB).

of the colonial world when it comes to defining borders. As different authors have shown (García-Álvarez and Puente-Lozano 2017; Guy 2008; Rankin 2008), the success of delimitation commissions in Europe relied both on sound knowledge of geographical conditions and on the response to conflicts of interest between local communities situated on either side of the border. By contrast, in the colonial context, expert commissions operated with less knowledge of the environment, with all of its complex topographic, linguistic, and cultural variations, and with far less regard for the needs and interests of the local society (Chester 2008; Donaldson 2008b; Schofield 2008; Seligmann 1995).

The French and Spanish governments had a great deal of experience in the demarcation of borders. Between 1853 and 1868, a joint commission made up of diplomats, and military officers had succeeded in delimiting the 685 km of the Pyrenean border, going from the Cantabrian Sea to the Mediterranean (Capdevila 2009). However, this experience would not be easily translated to the Moroccan territory. On November 27, 1912, Minister of State Manuel García Prieto and the French ambassador to Madrid, León Marcel Geoffray, signed the agreement that established the border between the French and Spanish protectorates in Morocco.[2] This agreement marked the end of a long process of negotiation on the partition of the Alawite Empire, dating back to 1900 (Pastor Garrigues 2013). Nevertheless, the border was but very briefly and confusingly described under the terms of the agreement, which had been drawn up without having a large-scale map of the area to hand (Urteaga 2006a, 2012). The most accurate Spanish map, which was to a scale of 1:500,000, had been prepared in 1904 and was a typical product of office cartography.[3] In the 1912 edition of this map, the line drawn was so thick that rather than define the border; it served to discreetly conceal its uncertain course (Fig. 9.1). French cartography was no better.

The 1912 agreement established an apparently easy path for the border at the eastern and western extremes of the territory, and a frankly complicated one elsewhere. At the eastern extreme, the border was to follow the course of the Moulouya River, from its mouth to 1 km downstream from Mechra Klila (Klila ford). Given its permanent flow, the Moulouya was an apparently easy boundary to establish. In the western area, the border had to follow the 35th parallel north, starting from the Atlantic Ocean until it reached the course of the Loukkos River, near Ksar el-Kebir. However, the extensive middle sector between the courses of the Loukkos and Moulouya rivers, with a linear path exceeding 300 km, was very poorly defined in the agreement, with confusing wording and a lack of any kind of reliable cartographic backing. Its demarcation would therefore be a source of problems.

Some of these problems were rather predictable, even without having to set foot in Africa. Firstly, there were no good maps of the area, and there was no geodetic network to support topographic work. Secondly, the territorial allocation agreed upon by the French and Spanish governments would break up the territory of some large *cabilas*,[4] including those of Beni Bouyahi and Mtalsa, two of the largest in northern Morocco. Thirdly, control of the territory by the Makhzen[5] was precarious, and the colonial administration had not even taken its first steps at the time. Other problems would however emerge once the cartographers set foot on the ground.

[2]'Convenio entre España y Francia, celebrado el 27 de noviembre de 1912, para precisar la situación respectiva de los dos países en relación al Imperio xerifiano', in *Diccionario de la Administración Española*. 1913 Appendix. Imprenta del Asilo de Huérfanos, Madrid, 1913, pp. 170–176.

[3]On how this map came about, see Urteaga (2006b).

[4]The term designates the Berber tribes of North Africa and the territory they occupied. During the Spanish Protectorate the *cabilas* were the basis of the territorial organization.

[5]The term Makhzen designated the government or supreme authority in Morocco, and also the ruling elite of the country.

Fig. 9.1 *Mapa de la parte Norte de Marruecos* [Map of the Northern part of Morocco] Elaborated by the Commission of the Army General Staff Corps. Scale 1:500.000. Without date yet published in 1912 (Talleres del Depósito de la Guerra) *Source* AHCCB

9.3 Organization of the Boundary Commission

It was agreed in the Spanish–French agreement of 1912 that in order to set the exact course of the borders of the protectorates, a technical commission would be appointed by the French and Spanish governments, comprised of an equal number of representatives from both countries. In carrying out its work, the commission could take into account not only topographic features but also local conditions. The reports drawn up by the commission would not have enforceable value until they had been ratified by the respective governments.

The Spanish–French Boundary Commission in Morocco[6] (hereafter referred to as the Boundary Commission) was created with remarkable speed and was fully operational by the spring of 1913. It was a joint body, in which both parties acted on equal terms. All of the commission's members were military officers. On the French side, members were appointed by the director of the Geographical Service of the Army; on the Spanish side, by the head of the *Depósito de la Guerra* (War Office), Colonel Pío Suárez-Inclán y González.

The French section was led by Captain Georges Perrier (1872–1946), who was in charge of political negotiations and of overseeing the work. It also comprised a captain responsible for geodetic work, two captains in charge of topographic work, a military doctor, and an interpreter,[7] as well as support staff for the field work.

The Spanish section of the commission was appointed on March 22, 1913. It was headed by the commander of the General Staff, Rafael Alfonso de Villagómez y Núñez (1875–1942), a cartographer who, as we shall see, would spend much of his professional career in the protectorate of Morocco. General Staff captains José María Baigorri Aguado, Manuel Laguillo Bonilla, and César Voyer Méndez were

[6]In France, it was officially referred to as the 'Commission Mixte de Délimitation Franco-Espagnole au Maroc'.

[7]ACEGCGE, C-6bis-III, 10. Estado Mayor Central. Servicio Geográfico del Ejército: *Comisión de Estado Mayor en Marruecos. Comisión mixta de delimitación Hispano-Francesa en Marruecos. Comisión Geográfica de Marruecos. Breves extractos de la documentación existente en el Archivo (desde el año 1881 a 1936)*, 1947.

appointed to work under him. Selection of the commission's members was undertaken carefully, taking into account the possibilities at the time. Alfonso de Villagómez, José María Baigorri, and Manuel Laguillo had experience of carrying out cartographic work in Moroccan territory (Urteaga 2006a, 2012)[8] César Voyer Méndez, the youngest of the commission's members, had been stationed in Algeria as an attaché to French military forces in 1911 and 1912.

The War Office was in charge of providing the commission with the material and support staff required, and of establishing the technical rules governing its work. In order to carry out the field work, a group of topographers from the *Brigada Obrera y Topográfica de Estado Mayor* (General Staff Topographic Brigade) was added to the Boundary Commission. The material delivered to the commission was carefully selected. A letter from the commander of the General Staff, Juan López Soler, who at the time was head of the geographical section of the War Office, to Commander Alfonso de Villagómez, explains why. According to López Soler, the Boundary Commission's material 'must be the best and most sophisticated, since the French have extremely modern material'.[9] The Spanish section was therefore equipped, among other materials, with a Jobin model prismatic astrolabe, a transportable antenna radio receiver with a 25-m pole, a sexagesimal 5 s theodolite, a chronograph, and Invar wire for measuring baselines.

Instructions for the operation of the Boundary Commission were approved on April 8, 1913. The unit was directly under the remit of the Ministry of War with regard to staff, the technical aspects of the works, and the designation of the areas to be delimited. In terms of the schedule of works, assistance and escort details, relations with French commission members, and ordinary incidents, it was answerable to the Spanish High Commissioner in Morocco.

According to the agreement between the French Geographical Service of the Army and the Spanish War Office, delimitation works were to commence at the westernmost section of the border, located near the mouth of the Loukkos River in the Atlantic Ocean. Rafael Alfonso de Villagómez and his men left for Larache in the first week of April 1913. On 15 April, they camped near the sanctuary of Sidi Suleiman, on the right bank of the Soueir River, about 24 km south of Larache. There, they were to meet the French commission members.

9.4 The First Attempt at Border Delimitation (1913–1914): Operations in the Loukkos Valley

The first border section to be delimited was that of the Loukkos region, in the protectorate's westernmost sector.[10] The November 1912 agreement offered a clear and apparently precise definition of the border in that sector. The border was to start from the valley of the Loukkos River, located between the *cabilas* of Sarsar and Tlig. 'From this point on', the agreement states, 'it will round Jebel Ghani, leaving this mountain in the Spanish zone, with the understanding that no permanent fortifications will be constructed there. Finally, the boundary will join the 35th parallel north, between the *douar*[11] of Mgaria and the *marya* of Sidi Slama and will run along this parallel to the sea'.[12] In short, a typical colonial border outline: largely a simple straight line running along a parallel over a short path, slightly over 40 km. However, things were not going to be that simple. Even boundaries that

[8]Urteaga 2006a, 2012. Unless otherwise stated, all translations of cited foreign-language material in this article are our own.

[9]ACEGCGE, C-6bis-III, 10. Estado Mayor Central. Servicio Geográfico del Ejército, op. cit., 1947, 73. Letter from Juan López Soler to Rafael Alfonso de Villagómez, 2 April 1913.

[10]This section is based on a previous work, which has been redrafted (Urteaga 2014).

[11]The *douar* is a small settlement of nomadic or semi-nomadic populations, and by extension the basic unit of rural communities.

[12]*Convenio fijando la respectiva situación de España y Francia en Marruecos, firmado en Madrid el 27 de noviembre de 1912*. Art. 2.

coincide with the path of a parallel must solve a geodetic problem, namely: Where is the parallel of reference actually located?

The two sections of the Joint Boundary Commission met at the Sidi Suleiman camp on April 16, 1913 to discuss a provisional working plan. The first issue addressed was precisely what form should be given to the delimitation along the 35th parallel north. A strictly geometric boundary may have made sense in a deserted area. However, the Loukkos region was populated, with many agricultural and cattle-raising settlements. Therefore, it was inevitable that a border in the form of a straight line would divide *cabilas*, fractions of *cabilas*, and even simple *douars*. After considering the problem, the commission members made a prudent decision: The course of the border had to accommodate the terrain's features and local conditions, and the parallel would serve as a reference to assess any compensations deemed appropriate in favor of either side.[13] At the same meeting, it was agreed that delimitation operations would start in the west, on the shores of the Atlantic Ocean, and would continue eastward without interruption. The two sections of the commission would perform their work independently, albeit establishing common survey markers for geodetic triangulation.

At another meeting, held a few days later, it was agreed to undertake the astronomical and geodetic work required to determine the 35th parallel north. Next, the topographic survey of a 4-km-wide strip along the parallel would be carried out, in order to be able to undertake the definitive study of the border. Field work began on 28 April. On the Spanish side, the astronomical and geodetic operations were directed by Commander Alfonso de Villagómez and Captain Manuel Laguillo Bonilla, who was familiar with the area. As for information activities, they were assisted by Luis Palanca Martínez-Fontán, captain of the Corps of Engineers, who was provisionally assigned to the Boundary Commission. Alfonso de Villagómez and Manuel Laguillo carried out the observation of a chain of triangles along the 35th parallel north, consisting of 21 markers and 22 triangles. In order to scale the triangulation network, joint observation of the Sidi bou Rziguine baseline was carried out with the French geodesists. This is a 5-km-long baseline situated south of the 35th parallel north, very close to the coast. Latitude, longitude, and azimuth were established at the northernmost point of the baseline, i.e., the closest to the coast. This work was captured in a document prepared by the French section of the Boundary Commission, which presents the full set of geodetic triangulations (Fig. 9.2).

After performing successive latitude calculations in order to re-establish the 35th parallel north, a major surprise arose: the aforementioned parallel was in fact located a fair distance further north than expected. Indeed, both French and Spanish maps traditionally placed this minor circle almost one degree to the south. This is apparent, to mention two maps published by official bodies, in the map *Maroc Septentrional d'après les documents les plus récents*, published in France by the Geographical Service of the Ministry of Foreign Affairs (1:500,000 scale, 1906), and in the *Mapa de la parte Norte de Marruecos*, published almost at the same time in Madrid by the War Office (1:500,000 scale, 1905). In both cases, the city of Ksar el-Kebir is clearly located north of the 35th parallel north.

However, the geodetic observations of the Joint Boundary Commission produced an unquestionable result: The 35th parallel ran slightly north of Ksar el-Kebir. Naturally, the error arose from the imprecision of previous measurements, which were made too hastily and without adequate instruments. Needless to say, the implications extended far beyond correcting a mere cartographic error. If the 35th parallel moved north, the border would straddle the city of Ksar el-Kebir, as well as the railway line under construction at the time, which was supposed to link this town with Larache. This was an unwelcome prospect for Spanish interests. Suddenly, what appeared straightforward became onerous. At any rate, the border could not run along a straight line.

[13]ACEGCGE, C-6bis-III, 10. Estado Mayor Central. Servicio Geográfico del Ejército, op. cit., 1947, p. 74.

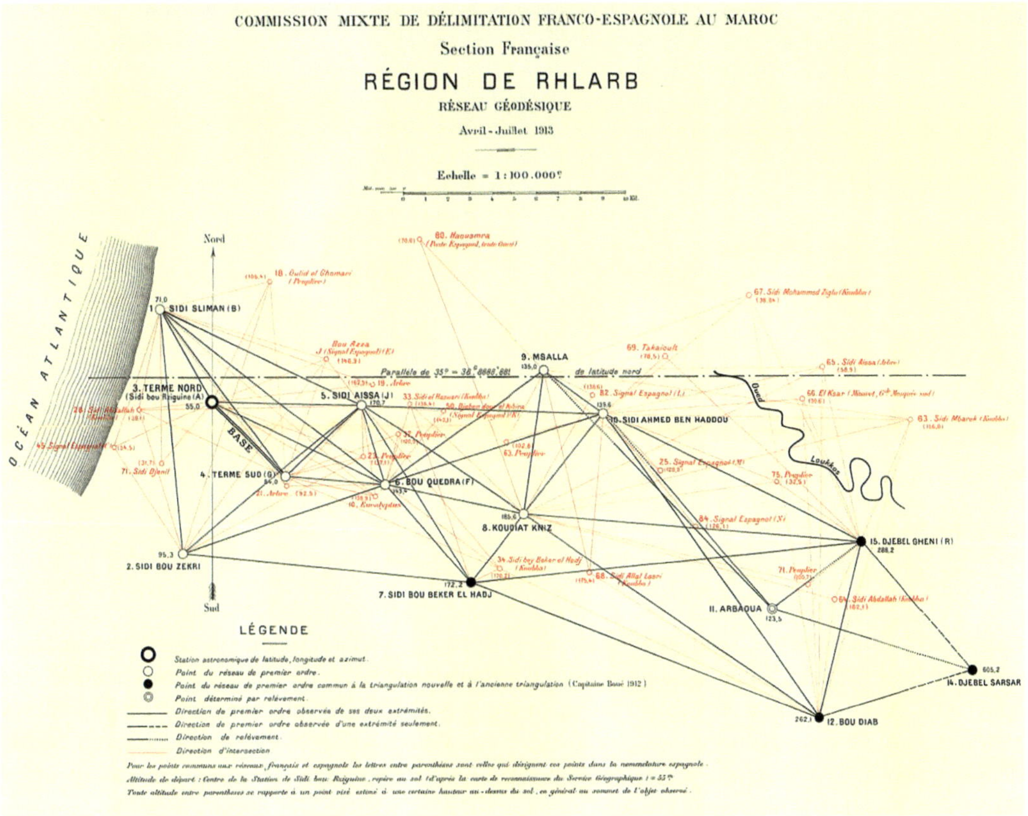

Fig. 9.2 *Région de Rhlarb. Réseau Géodésique. Avril-Juillet, 1913*. [Region of Rhlarb. Geodesic network. April–July, 1913] Scale 1:100.000. Franco-Spanish Joint Boundary Commission of Morocco. French Section. In the points being common to both the French and the Spanish triangulation, the Spanish toponymy is indicated in parenthesis. *Source* IGNM, CCBY 4.0 ign.es

On June 5, 1913, Alfonso de Villagómez sent a report to the head of the War Office raising the issue.[14] If the border was to be shifted at Ksar el-Kebir, they would need to offer the French more land in other areas. The head of the Boundary Commission suggested handing over the fraction of Oulad Skhar, located to the west. Examining land compensations would require many hours of negotiation and the involvement of diplomats in Madrid and Paris. While these negotiations were underway, the cartographic work continued on its course.

The topographic survey of the border strip was entrusted to captains José María Baigorri and César Voyer Méndez, who led a group of military topographers from the General Staff Topographic Brigade. Topographic work was carried out at a scale of 1:50,000, in a west–east direction, over a strip of variable width that ended up covering approximately 150 km^2 (Fig. 9.3). For the area between Bou Lehiba and the marabout of Sidi Mohammed Ben Hadou, a more detailed survey was carried out, at a scale of 1:20,000. During the redefinition of the border, nine fractions (subdivisions of *cabilas*) were demarcated, with a total perimeter of more than 200 km.

Delimitation operations were carried out without issues during the months of April and May 1913. However, the field work met with difficulties beginning in 4 June, from which date various positions in the area were shot by local rebels. The topographic parties had to

[14]ACEGCGE, C-6bis-III, 10. Estado Mayor Central. Servicio Geográfico del Ejército, op. cit., 1947, p. 75.

Fig. 9.3 Topographic survey of the border area from the sea to Lucus region. Morocco Boundary Commission. May 1913. Scale 1:50.000. Four color handwritten map on Canson paper. Relief using contour lines. One sheet of 95 × 71 cm. *Source* ACEGCGE. MAR–C.4–066. Biblioteca Virtual del Ministerio de Defensa BMDB 20,149,015,863

temporarily withdraw: the Spanish to Ksar el-Kebir and the French to Arbaoua. On 8 June, the topographic work resumed, only to be yet again interrupted on 23 June due to a shooting at the Mensah camp.[15] Topographic operations finally came to an end in the first week of July, upon reaching the Mount Ghani area, and the Spanish section of the Boundary Commission was able to return to Madrid.

After devoting a couple of months to office work, Rafael Alfonso de Villagómez and César Voyer Méndez traveled to San Sebastián on September 18, 1913 to meet with the French section. On 21 September, the delimitation report for the section between the Loukkos River and the sea was signed into force.

The area controlled by the Spanish troops in the west of the protectorate, albeit precariously, ended almost at the very gates of Ksar el-Kebir. To the east, as soon as one entered the territory of the *cabila* of Ahl Serif, one ventured into an unsubdued area, where there was no possibility of carrying out normal topographic work. The ambition of pursuing delimitation operations from the Atlantic to the east without interruption had to be abandoned. At the meeting between the Spanish and French delegations of the Joint Boundary Commission in San Sebastián in September 1913, an agreement was made to meet again at the end of January 1914, in order to discuss the border in the section between the ford of Klila (Mechra Klila) and the mouth of the Moulouya River—in other words, the border in the protectorate's easternmost part. This was a river section of about 80 km in length, which could be defined without too many difficulties, since the Moulouya is a perennial river.

Delimitation work took place in the Moulouya basin from late November 1913 to April 1914. The information that I have been able to gather regarding the course of these operations is piecemeal and incomplete. I have not been able to locate the geodetic triangulation for the area, nor any evidence of the cartographic work carried out. In any event, the available documentation makes it possible to establish a fairly complete account of the activities of the Spanish section of the Boundary Commission.

Following orders from the head of the War Office, Commander Rafael Alfonso de Villagómez left for Melilla on November 20, 1913. He was accompanied by General Staff captains José María Baigorri Aguado, César Voyer Méndez, and José Centaño de la Paz: the latter replacing Captain Manuel Laguillo Bonilla, who was assigned to the General Command of

[15]AGMS, Leg. A-1.228.

Larache.[16] This time, we have almost full knowledge of the support contingent from the General Staff Topographic Brigade, which would be in charge of the topographic work. It was made up of nine workshop masters and thirty assistant topographic workers.

Alfonso de Villagómez divided his staff into two sections, one dedicated to geodetic work and the other to topography. The topographic section, in charge of surveying the course of the Moulouya from its mouth, was under the command of Baigorri Aguado. It was made up of the three General Staff captains, seven workshop masters, and twenty workers. In order to carry out their work, a request was made to the General Commander of Melilla to provide four tents, required to accommodate staff, as well as four water tanks and six mules for transportation.[17] The geodetic section, which was made up of the rest of the commission's staff, came under the direct control of Commander Villagómez. It was tasked with primary triangulation, astronomical calculations, and measuring a baseline in the Selouane highway. Given that geodetic triangulation required the building of permanent signs, support had to be requested from the Engineers' Command, along with five additional mules for transporting impedimenta. Both the geodetic and topographic sections relied on a Mía (company) of the Indigenous Police as their escort force while carrying out the field work.

Alfonso de Villagómez began observations in Melilla, in the Alfonso XII fort, in order to determine its geographical coordinates using the prismatic astrolabe. On December 30, 1913, he wrote to Madrid stating that he was temporarily suspending the astronomical task until he could make use of a radiotelegraph station enabling him to calculate the longitude.[18] In January 1914, the staff under his command began triangulation estimates. By March, a request was made to Madrid to urgently send baseline measurement equipment to conduct the observation of the Selouane–Taouima geodetic baseline, which was to be used to guide and scale the triangulation of the area.

Meanwhile, the section commanded by Baigorri Aguado began the topographic survey from the mouth of the Moulouya, using the Cabo de Agua camp as an operational base. From here, the topographers advanced upstream, in parallel to the Kebdana mountain range, until they reached the Saf Saf ford, very close to the Zaio camp.

The topographic survey of the Moulouya basin between Cabo de Agua and Mechra el Melha at a scale of 1:50,000 was completed on March 16, 1914, and observation of the Taouima baseline ended on 15 April. Almost immediately, the order was received that the commission's staff should go to Madrid to carry out office work. On July 15, 1914, once this work was complete, the head of the commission left for Paris on a secondment, in order to hold negotiations with the French section of the Joint Boundary Commission. The circumstances could not have been less favorable. Negotiations were abruptly suspended on 9 August due to the outbreak of the First World War.

As was to be expected, the problem of delimiting the protectorate of Morocco vanished from France's agenda for five long years. Absorbed in the immense effort of the First World War, the French army was forced to interrupt non-essential activities—among others, those relating to the delimitation of borders. When attempts were made to resume these works five years later, diplomatic relations between France and Spain had undergone a considerable shift, having gone from collaboration to secret distrust (Urteaga and Nadal 2010).

9.5 The Second Attempt (1919–1921): Topographic Operations in the Guerrouaou Plateau

Condemned to a standstill by the war, the Boundary Commission may well have been dissolved. However, the High Commissioner in

[16] AGMS, Section CG, Leg. L-16.

[17] AGMM, Africa Collection, Box 702, Leg. 124. Letter from the Head of the Boundary Commission to the General Commander of Melilla, 3 December 1913.

[18] ACEGCGE, C-6bis-III, 10. Estado Mayor Central. Servicio Geográfico del Ejército, op. cit., 1947, 76.

Morocco was not willing to get rid of a cartographic unit with experience of field work in Moroccan territory and well-endowed in terms of staff and means. Consequently, General Francisco Gómez Jordana, High Commissioner in Morocco, and Colonel Pío Suárez-Inclán, head of the War Office, agreed to keep the Boundary Commission active, and to assign its staff to carrying out support tasks for the Geographical Commission of Morocco, which oversaw the entire protectorate's cartography. In keeping with this purpose, the Boundary Commission, which remained under the authority of Commander Alonso de Villagómez, was sent to the Ghomara region to carry out surveying for the Protectorate's Military Map, at a scale of 1:100,000 (western sector).

They were relieved of this task in 1919 by the new High Commissioner, General Dámaso Berenguer, who led an ambitious plan for military control over the entire territory of the protectorate. Over the course of the First World War, diplomatic relations between France and Spain had been strained by the presence in the Spanish zone of German agents involved in arms sales, as well as the emergence of sporadic conflicts in some border *cabilas* (Woolman 1971). One of the main sources of conflict was the extensive territory of the *cabila* of Beni Bouyahi, located in the Guerrouaou basin, east of the Rif mountain range and straddling the border between the protectorates of France and Spain.[19] Said *cabila* was made up of eight fractions, three located in the French zone, two in the Spanish zone, and the remaining three straddling the border. Given that the Beni Bouyahi was nomadic shepherds, establishing their water supply areas and grazing spaces was a key task when carrying out a demarcation of the border that would be acceptable to both colonial authorities and the indigenous population.

In these circumstances, Dámaso Berenguer decided to reorganize the Boundary Commission and return it to its original role. On September 10, 1919, he summoned Commander Alfonso de Villagómez to explain the situation in the Guerrouaou border area and issue him with instructions. Given the situation of distrust between the French and Spanish governments, it was not a case of immediately commencing demarcation operations. Rather, the intention was to carry out the exploratory work required to clear up the border dispute. Once this tentative phase was over, 'formal delimitation work agreed between both governments [could] begin'.[20]

Alonso de Villagómez received the order to leave immediately for Melilla, accompanied by Captain Isidoro González Adalid, two military topographers, and fourteen topographic assistants belonging to the General Staff's Topographic Brigade. This time, the Boundary Commission was strengthened by the addition of an officer from the Indigenous Police forces, Captain José Pomés Damont, belonging to the Mía (company) of Guerrouaou. This officer was later replaced by Captain Jesús Jiménez Ortoneda. The temporary addition to the General Staff units of seasoned members from other branches or corps with good knowledge of the terrain had been a common practice almost from the very start of cartographic surveys in Morocco (Urteaga 2006). And the reasons were almost always the same: difficulties relating to information work in particularly complex social and environmental surroundings, as well as linguistic competence. In the years under study, the protectorate had neither a gazetteer nor a standardized way of conveying Berber phonetics in Spanish. In the eastern region of the Rif, which was the area chosen for the Boundary Commission's new activities, various dialects of Shilha were spoken, in addition to Arabic, and topographic features had different names depending on the different *cabilas*. Establishing place names was, under these circumstances, a particularly difficult task. On the other hand, the Boundary Commission's cartographers were forced to maintain close contact with the heads of the *cabilas* and

[19] *APJO. International Boundary Commission of Morocco. Notes on the border outline. Beni Bouyahi sector. Typescript, 1929.*

[20] AGMM. Africa Collection, Box 260. The High Commissioner of Spain in Morocco to the Minister of War, 10 September 1919.

fractions in which the surveys were being carried out. I am unaware of Pomés Damont's linguistic background, but the choice of Captain Jiménez Ortoneda to provide support to the Boundary Commission is hardly surprising; at the time, Jiménez Ortoneda had been in Morocco for almost fifteen years, spoke Arabic and Shilha, and was in charge of the Afso information bureau, in the *cabila* of Beni Bouyahi (Villanova and Urteaga 2009).

The members of the Boundary Commission set off for Melilla on October 28, 1919, aboard the *Delfín* steamboat.[21] From Melilla, they immediately left for Afso, where they set up their base camp for the field work. This move by the Spanish cartographers into the Guerrouaou plateau was met in kind by an almost simultaneous move by the French military cartographers acting under the authority of the Topographic Bureau of Morocco.

On 24 November, the Spanish were made aware of the arrival in Afso of French captains Mege and Tavernier from the Topographic Bureau, 'escorted by 50 horses', with the aim of mapping the Guerrouaou.[22] Following the verbal instructions he had received, Alfonso de Villagómez negotiated an amicable division of tasks. The Spanish commission would provide French cartographers with data from the Kerker (1004 m), and Beni Idder survey markers located along the line of peaks at the eastern end of the Rif mountain range. The Spanish military would work to the north of the aforementioned markers, and the French would work on the southern side, with a view to sharing their results once each party had concluded its operations.

A week later, the High Commissioner gave his approval to this agreement, which was informal in nature, in the following terms: 'I approve the amicable agreement for topographic work in the Guerrouaou established between Commander Villagómez and French officers, which facilitates work in common and will in due course shorten the work of the official commissions appointed by the respective governments'.[23]

Field work must have started right away. Unfortunately, it has not been possible to find either the topographic triangulation or the backfill work carried out in the Guerrouaou. However, the surviving documentation allows us to confirm that topographic operations were ongoing throughout 1920, and that successive follow-up meetings were held between the Spanish commission members and the French cartographers. On 8 January of that year, in Afso, Commander Alfonso de Villagómez received Commander Cornudet, from the Geographical Service of Morocco (since 1920, the successor to the Topographic Bureau), in order to check adjustments to the shared geodetic markers. A month later, Villagómez was promoted to lieutenant colonel. In theory, this promotion should have implied a new posting, but the High Commissioner immediately responded by requesting that he remain in his role within the commission, since 'his discharge would cause extraordinary disruption, especially in the present times, when he is carrying out highly sensitive tasks in the disputed region of the Guerrouaou'.[24] On June 9, 1920, Commander Édouard de Martonne, head of the French section, arrived in Afso for another meeting with Spanish cartographers, and on 2 December, Alfonso de Villagómez traveled to Hasi Ouenga, in the French protectorate, to meet with Captain Tavernier and to visit the defenses that were being built in the area.

This outing into the French zone was practically Alfonso de Villagómez's last act as head of the Boundary Commission. Shortly afterward, it was taken over by Lieutenant Colonel José Molina Cádiz (1869–1926), a seasoned cartographer with extensive experience of the territory from the Melilla General Command, with two young General Staff captains under his command: Sigifredo Sainz Gutiérrez and Enrique Sánchez-Monge y Cruz. From the former team

[21]AGMM, Africa Collection, Box 260. Radiogram to the ministries of war and state, 30 October 1919.

[22]AGMM, Africa Collection, Box 260. From the General Commander of Melilla to General Berenguer, 24 November 1919.

[23]AGMM, Africa Collection, Box 260. From the High Commissioner to the General Commander of Melilla, 4 December 1919.

[24]AGMM, Africa Collection, Box 260. Letter to the Ministry of War, 13 February 1920.

that had worked with Villagómez, only Captain Jiménez Ortoneda remained in the Guerrouaou.

It is difficult to ascertain whether the dispute over the Guerrouaou remained a priority at the beginning of 1921. Since the end of January of that year, the Ministry of State had been demanding from the Ministry of War that the staff of the Boundary Commission be expanded, and that part of said staff be assigned to the western area of the protectorate, in order to work in the border region between the Loukkos and Ouergha rivers. It is quite likely that, by then, the priorities in the territory of Melilla were already different. The spring of 1921 was a period of optimism, almost euphoria, in the Melilla Command. Troop movements ordered by General Manuel Fernández Silvestre had extended the occupied area from the course of the Kert River to the banks of the Amekran River, 50 km to the west, and very close to the *cabila* of Beni Ouriaghel considered the heart of resistance in the Rif. The plan to complete the total occupation of the protectorate then seemed within reach. However, this was completely turned on its head within just a few months. Between July 21 and August 9, 1921, the Spanish army in Morocco suffered a tragic defeat at the Battle of Annual, known in Spain as the Disaster of Annual, which led to the loss of thousands of lives and ended up endangering the Melilla Command as a whole. The Disaster of Annual dealt a square blow to the Boundary Commission, which was left severely weakened. Captain Sánchez-Monge died defending Monte Arruit on August 9, 1921. Fellow captain Sainz Gutiérrez was taken prisoner in the same place and would spend a long time in captivity. As may be expected, border demarcation work once again came to a halt following the Disaster of Annual. Five long years would go by before it could resume.

9.6 The Third Attempt (1926–1931): Topographic Work in the Foothills of the Rif

The unsteady and unpredictable course of the Rif War ended up forcing a new entente between France and Spain, so as to put an end to Abd el-Krim's insurrection. At the same time, however, the course of military operations further complicated the problem of territorial demarcation. Indeed, the border that had been agreed to in 1912 was modified de facto at a number of points during the Rif War from 1921 to 1926. During the conflict, French troops penetrated the Spanish zone and established defensive positions north of the Ouergha River. This occupation, initially intended to be temporary, acquired a more permanent standing following the agreement signed between the French and Spanish governments in Madrid on June 25, 1925.[25] This agreement left the *cabilas* of Marnissa, Beni Bouslama, Fenassa, Beni Ouensel, Metioua, and part of the *cabila* of Beni Zeroual, within the French zone. During the campaign that followed the Al Hoceima landings in August 1925, new points of dispute arose in the *cabilas* of Beni Bouyahi, Gueznaia, and Beni Zeroual, in which the French army secured and maintained advanced positions in areas claimed by Spain.

In an attempt to resolve these border issues, a new diplomatic agreement was negotiated and signed in Paris on July 10, 1926, calling for the re-establishment of the technical boundary commission provided for in the Spanish–French agreement of November 27, 1912. By virtue of this agreement, the decision was made to resume demarcation work, starting with the border sector that stretched from the Moulouya River to Jebel Rokdi, a sector that was occupied and had been pacified.[26] On August 26, 1926, the Ministry of War set about reorganizing the Boundary Commission for the resumption of demarcation activities.[27] The unit was considerably strengthened and was placed under the command of

[25]AMAEC, Restored Collection. Leg. 3.846, Exp. 27. Agreement of 25 July 1925 between Spain and France with regard to the boundary of the two zones of Spanish and French influence in Morocco.

[26]AMAEC, Restored Collection. Leg. 3.846, Exp. 27. Agreement regarding the pacification of Morocco and the establishment of a regime in the confines between the areas of Spanish and French influence, 10 July 1926.

[27]Circular strengthening the Boundary Commission of Morocco, *Diario oficial del Ministerio de la Guerra*, Year XXXIX, nº 190, 26 August 1926, volume III, p. 626.

Lieutenant Colonel Antonio Aranda Mata (1888–1979), a prestigious military officer with sound knowledge of the terrain, who had been in charge of geographical works in Morocco since 1923. Appointed to be under the command of Aranda Mata was General Staff commanders Emilio Esteban Infantes and José Figuerola Alamá, a captain from the same corps, Rafael Gómez Redondo, and once again, Jesús Jiménez Ortoneda, the multilingual indigenous affairs officer, who had attained the position of commander.

In mid-October 1926, Antonio Aranda Mata traveled to Rabat to meet with Lieutenant Colonel Charles Octave Chicoyneau de Lavalette (1872–1963), head of the French section of the Boundary Commission. During the meeting, which took place from 11 to 13 October, the main technical aspects of the border survey were agreed upon. The first task was to perform a first-order triangulation, which setting off from the Midar astronomical baseline, would stretch from Mechra Klila to the Ouergha River. The outer line of markers for this network would be common to the French and Spanish works; once the values of the geographical coordinates, azimuths and marker altitudes were verified. The second task involved carrying out a topographic survey along the likely border, with an average width of 6 km, which could be widened as much as required, wherever any differences of opinion arose regarding the outline. Thirdly, in parallel with these tasks, an agreement would be sought on the course and route of the Moulouya River, from its mouth to Mechra Klila, which had been surveyed by the Spanish commission members in 1914. Geodetic operations were to begin immediately, while topographic activities would commence 'as soon as the rest of the staff from the French commission, who are still in France, arrive in Morocco'.[28] As we shall see, things did not quite turn out in keeping with these projections.

Autumn 1926 was cold and rainy. Storms made communications difficult and delayed field work. Most of the peaks that were to serve as geodetic markers were covered in snow. At the end of November, a meeting with Théodore Steeg, Resident-General of France in Morocco, dampened the mood. As announced by Steeg, given the weather conditions, the French section of the Boundary Commission would not start topographic work until the spring of 1927, and in the time being, they would focus on making whatever progress was possible in geodetic and office-based work.[29]

Despite this setback, the Spanish section did not grind to a halt. In the final months of 1926, the cartographers under the orders of Aranda Mata prepared a 5-km-long geodetic baseline between Midar and Boufarkouch, completed the project to triangulate the terrain between Mechra Klila and Jebel Arzou, built the Spanish zone's first-order markers, and explored the border area through the *cabilas* of Beni Bouyahi, Mtalsa, and Gueznaia.[30]

From January to March 1927, the geodesist Lieutenant Rainoird, from the French section of the Boundary Commission, carried out second-order triangulation work in the *cabilas* of Beni Bouyahi and Mtalsa, in order to secure the geodetic link with the triangulation carried out by the Spanish section. Once this task was complete, the two sections of the Boundary Commission met in Melilla on 31 March to compare the network's common values. Over the course of the meeting, the Spanish delegation sought to get the topographic work underway immediately. The French delegation, however, objected to this, citing 'staffing difficulties'. In his report to the Director General of Morocco and the Colonies, Lieutenant Colonel Aranda Mata expressed his disappointment, noting that he had observed 'a calculated resistance to speeding up activities, especially topographic work'.[31]

[28]AGAH. Morocco Section, Box 81/10.013. Aranda Mata to the Director-General of Morocco and the Colonies, 17 October 1926.

[29]AGAH. Morocco Section, Box 81/9.992. From the High Commissioner to the President of the Council of Ministers, 28 November 1926.

[30]AGAH. Morocco Section, Box 81/10.013. Aranda Mata to the Director-General of Morocco and the Colonies, 30 November 1926.

[31]AGAH. Morocco Section, Box 81/10.013. Aranda Mata to the Director-General of Morocco and the Colonies, 31 March 1927.

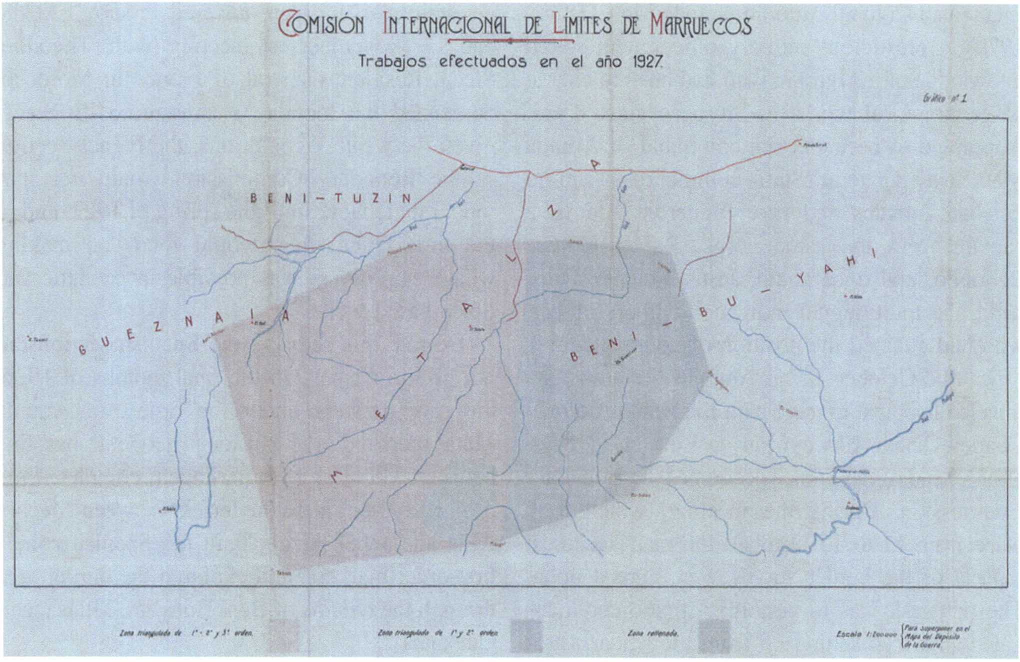

Fig. 9.4 *Trabajos efectuados en el año 1927* [Works undertaken in 1927]. Graphic no. 1. Morocco International Boundary Commission. Scale 1:200.000. Manuscript on greaseproof paper, 88 × 44 cm. *Source* Ministerio de Cultura y Deporte [Spain], AGAH. Fondo Dirección General de Marruecos y Colonias, IDD (15) 003.001, box 81/10013

After several months of delay, the topographic survey finally got underway on May 20, 1927. Work began on the Tendri plateau (the Oulad Taleb to Ain Zohra fraction, *cabila* of Mtalsa), 'in order to gather the required information as soon as possible to end the constant border incidents that have been taking place in this region'.[32] The common survey area, as agreed, was 6 km wide, 3 km on either side of the likely border. The markers for third-order triangulation were set in the course of operations, and the backfill work was carried out through the system of tachymeter itineraries. The backfill work was carried out at a scale of 1:20,000, with an average of 40 points per square kilometer in populated areas and about 25 elsewhere. From the initial area of operations in Mtalsa, work spread toward the territory of the Beni Bouyahi (see Fig. 9.4).

From mid-1927 onward, an increasingly disappointed Aranda Mata came to report what, in his view, were delaying tactics by the French authorities to postpone the execution of the survey as much as possible. Firstly, the Resident-General forced both sections of the Boundary Commission to work together, and with the same number of field work parties. In practice, this meant closely harmonizing the schedule of operations and leaving the initiative for its progress in the hands of the French. Once July came, topographic work had to be suspended due to the withdrawal of the French section, 'using heat as a pretext, which, although strong, did not in any way prevent work from continuing'.[33] During the second quarter of 1927, Spanish cartographers were stuck doing office work. They had planned to meet the French side in December in Tétouan to compare the work carried out throughout the

[32]AGAH. Morocco Section, Box 81/10.013. Aranda Mata to the Director-General of Morocco and the Colonies, 1 June 1927.

[33]AGAH. Morocco Section, Box 81/10.013. Aranda Mata to the Director-General of Morocco and the Colonies, 31 July 1927.

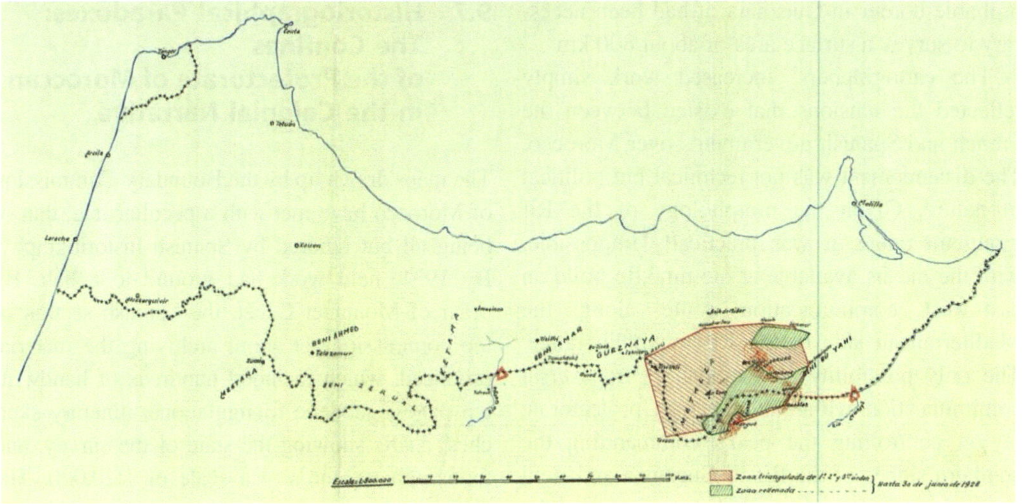

Fig. 9.5 *Zona triangulada de 1º, 2º y 3º orden. Zona rellenada. Hasta 30 de junio de 1928*. [First, second and third-order triangulated zone. Filled zone. Until 30 June 1928] Works undertaken by the Boundary Commission until 30 June 1928. Morocco International Boundary Commission. Scale 1:800.000. Manuscript on greaseproof paper, 48 × 24 cm. *Source* Ministerio de Cultura y Deporte [Spain]. AGAH. Fondo Dirección General de Marruecos y Colonias, IDD (15) 003.001, box 81/10013

year, but the meeting had to be suspended 'because their boss gave the excuse that he had sprained his foot'. By then, it was already clear that the initial momentum had been lost.

In January 1928, Lieutenant Colonel Federico Montaner Canet and Commander Darío Gazapo Valdés joined the Spanish section of the Boundary Commission, replacing Emilio Esteban Infantes and José Figuerola Alamá. Field work resumed in the spring of that year in the *cabila* of Beni Bouyahi. However, operations would be considerably slower and more difficult than expected. The Spanish cartographers initially tried to limit their work to the 6-km-wide strip straddling the border, as per the 1912 agreement. By contrast, the French section of the Boundary Commission demanded that the border line agreed in 1925 also be mapped, as well as the de facto border since 1926. Following a cumbersome negotiation, it was agreed that the survey would cover the entire stretch of land between the line of the 1912 agreement, and the de facto border established at the end of the military operations. Consequently, a much larger survey had to be carried out than expected, without making satisfactory progress toward the goal of demarcating the border. The graphic report of the work carried out in 1928 clearly reflects the surprising enlargement of the area surveyed, to more than 40 km in width, and its shift to the north, as proposed by the French authorities (Fig. 9.5).

The next year, topographic work began at the boundary between the *cabilas* of Mtalsa and Gueznaia, progressing westwards from there. The border in this area should have been clear in principle, since it theoretically followed the dividing line between the Kert and Nekor river basins to the north, and those of the Inaouen and the Ouergha to the south. In practice, however, the opposite was the case. In the course of the war, the French army had established positions in Izkritene, Boured, and Tamchecht, which allowed it to have control over the line of peaks, and with it the entire territory of Gueznaia. As a result, the French expected the working focus for the survey to be the northern boundary of said *cabila*, thus forcing a fresh expansion of the mapped area. On March 28, 1930, Lieutenant Colonel Federico Montaner Canet (1874–1938), the new head of the Boundary Commission reported that in order to cover 35 linear km of probable border in Gueznaia; it had been necessary to survey a surface area of about 800 km^2.[34]

The cartographers' increased work simply reflected the tensions that existed between the French and Spanish governments over Morocco. The disagreement was not technical but political in nature. Given the morphology of the Rif mountain range, it was practically impossible, with the means available at the time, to build an east–west communication route along the Mediterranean side of the Rif mountain range. The only possibility of establishing transversal communication within the Spanish protectorate lay in controlling the peaks commanding the southern slopes of the Rif in Gueznaia and Beni Zeroual. Accordingly, the expansion of the French zone to the north of the Ouergha implied far more than a significant territorial loss for Spanish interests; it ruined the possibility of establishing efficient military communication from Melilla to Chaouen.

Spanish diplomatic efforts to redirect the situation, which had been ongoing since 1930,[35] met with France's unwillingness to relinquish a comfortable and advantageous strategic situation. French military occupation in the *cabilas* of Beni Bouyahi, Gueznaia, and Beni Zeroual continued, with diplomacy unable to resolve the dispute. It was therefore not possible to proceed with the full demarcation of the border. Nor was it even possible to continue with topographic work on the border to the west of Jebel Rokdi. The last graphic document to give an account of the state of topographic work in the border region reveals a very uncomfortable reality: Delimitation disagreements on the ground had led not to the demarcation of a border but to the drawing up of two divergent boundary lines in the foothills of the Rif (Fig. 9.6).

9.7 Historiographical Paradoxes: The Confines of the Protectorate of Moroccan in the Colonial Narrative

The maps drawn up by the Boundary Commission of Morocco have met with a peculiar fate: that of being all but ignored by Spanish historiography. By 1931, field work had ground to a halt. By order of Montaner Canet, the Spanish section of the commission set about archiving the material produced, which included hundreds of handwritten pages: geodetic triangulations, itinerary sketches, charts showing the state of the survey, and topographic records at a scale of 1:20,000. The topographic records were used in the drafting of the southernmost sheets of the Topographic Map of the Spanish protectorate of Morocco at a scale of 1:50,000, published by the War Office between 1927 and 1936 (Nadal et al. 2000). The remaining material was sent to the relevant civil and military archives. They have remained there to this day: handwritten and unknown.

There is, nevertheless, one notable exception. One of the maps designed by the Boundary Commission was indeed published and widely distributed. It is rather nondescript original title is *Gráfico político: Trazado fronterizo de los diversos tratados* [Political Chart: Border Outlines of the Various Treaties]. It is dated March 1932, in Melilla, bears the signature of Commander Darío Gazapo Valdés, the approval of Lieutenant Colonel Federico Montaner Canet, and the seal of the Spanish–French Boundary Commission of Morocco, Spanish Section. The manuscript map, which was originally drawn up at a scale of 1:500,000, was attached to a report prepared by Montaner Canet, addressed to Luciano López Ferrer, the High Commissioner of Spain in Morocco, summarizing the activities carried out by the Boundary Commission, together with some thoughts of a political–military nature.[36] López Ferrer was the first High Commissioner appointed

[34]AGAH. Africa Collection, Box 81/10.013. Report of the Spanish Section of the Spanish–French Boundary Commission of Morocco, 28 March 1930.

[35]AMAEC, Restored Collection. Leg. 394, Exp. 9. Note from the Director-General of Morocco and the Colonies, Dámaso Berenguer, to the Spanish Ambassador in Paris, on the continuation of the work of the Boundary Commission in Morocco, 14 May 1930.

[36]AMAEC. Restored Collection. Leg. 3.846, Exp. 27. Report submitted to His Excellency, the High Commissioner of Spain in Morocco, by the Boundary Section of the Geographical Commission of Morocco. Ceuta, April 1932.

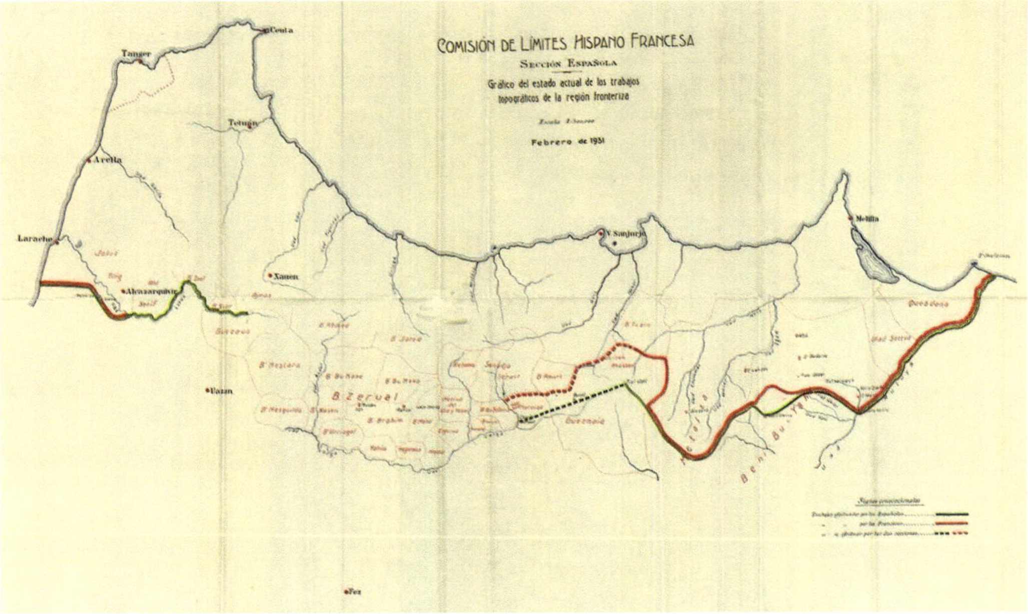

Fig. 9.6 *Gráfico del estado actual de los trabajos topográficos de la región fronteriza*. [Graphic of the current state of the topographical surveys of the border region]. Spanish–French Boundary Commission. Spanish Section. Scale 1:500.000. February 1931. Manuscript on greaseproof paper, 78 × 53 cm. *Source* Ministerio de Cultura y Deporte [Spain]. AGAH. Fondo Dirección General de Marruecos y Colonias, IDD (15) 003.001, box 81/10013

by the Republican government, as well as the first civilian to have held the position since 1923. The map that Montaner Canet offered him developed a political argument made to measure, namely the image of sustained frustration that demanded a vigorous diplomatic response.

The map represents the border lines supposedly agreed between the French and Spanish governments in 1902, 1904, 1912, and 1925, and the de facto boundaries established in 1926. The idea that can be deduced from it is as plain as it is categorical: France had gone back on the signed commitments, and to its own benefit succeeded in progressively reducing the territory assigned to the Spanish protectorate in Morocco. As with any powerful cartographic image, it skillfully combines facts and interpretations, certainties and musings (Fig. 9.7).

The southernmost line on the map draws out the southern boundary of the area offered to Spain in the exploratory talks of 1901–1902. French diplomacy at the time offered a partition of Morocco that would hand Spain the northern part of the Moroccan Empire, to the north of the Sebou River, including the cities of Fez, Taza, and Ouazzane. France, for its part, reserved for itself the heart of the empire, from the Sebou River to Agadir. The aforementioned negotiations did not come to fruition. Aware of its weakness on the international stage, the Spanish government did not dare to take the step of agreeing on the partition of Morocco. As such, the 1902 boundary does not refer to a border based on a treaty, nor even on a draft treaty; rather, it is pure imperial fancy, a treat offered to the Spanish government as part of the great tactical game of the partition of Africa.

During the summer of 1904, Spanish and French diplomats negotiated an agreement on the distribution of areas of influence in the Maghreb, which in practice established the partition of Moroccan territory between the two countries. In this round of negotiations, France reduced the offer it had put forward three years earlier, in particular the possible cession to Spain of the city of Fez and the Taza corridor. The Spanish

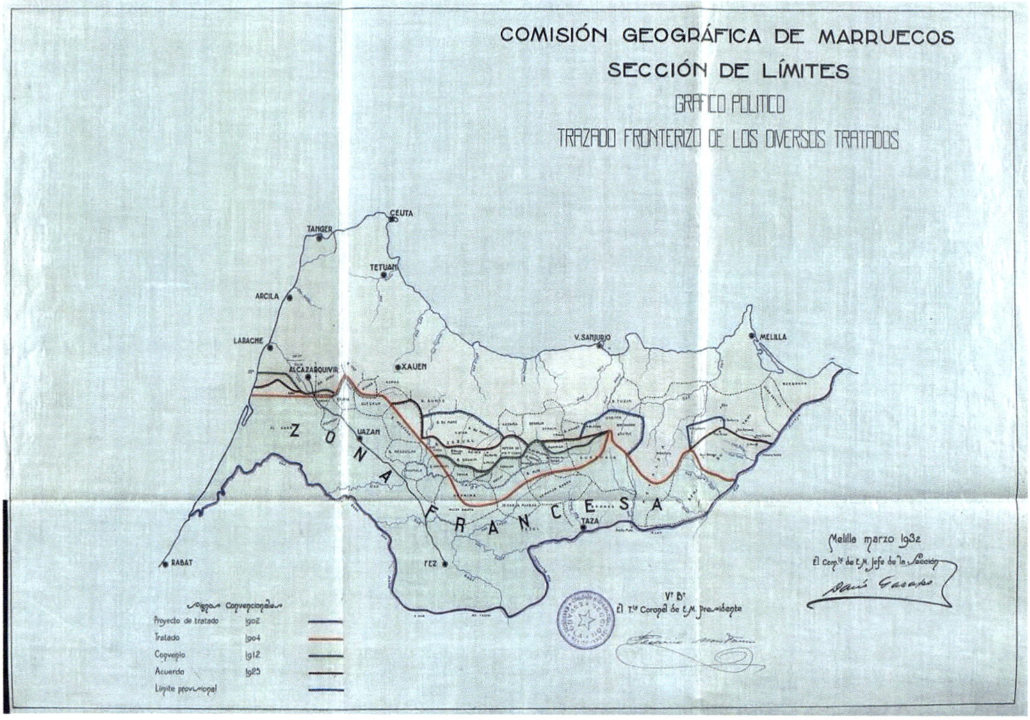

Fig. 9.7 *Gráfico político. Trazado fronterizo de los diversos tratados* [Political graphic. Boundary layout of the different treaties]. Composed by the Boundary Section of the Morocco Geographic Commission. Melilla, March 1932. Manuscript on greaseproof paper, 65 × 41 cm. No scale. [1:1.000.000]. The graphic represents boundary layouts as agreed upon in 1904, 1912, and 1925, and the limits actually established as of 1926. *Source* Ministerio de Cultura y Deporte [Spain]. Fondo Archivo Renovado del Ministerio de Asuntos Exteriores, IDD (10) 000.000, box 82/10524, exp. [file] 27

government ended up agreeing, settling for a narrow territorial strip of just over 20,000 km^2 in northern Morocco, from which the city of Tangier and its *alfoz* (territory) were kept out. The Spanish–French agreement of 1904 was not a treaty either, in the sense usually given to this term in the field of international relations. It was never conveyed to the Sultan, or to the Moroccan government, and it was kept secret until 1912. As for the border agreements of 1912 and 1925, and the geographical uncertainties surrounding their implementation on the ground, this has already been discussed in the preceding pages, and we need not dwell on it. Their outlines were as decisive on paper as they were volatile in reality.

However, the volatility of borders can be good fuel for nationalist rhetoric. The map of the Boundary Commission was redesigned and published without specifying its authorship in the *Geografía de Marruecos* [Geography of Morocco] prepared by the Historical Commission of the Moroccan Campaigns, published in 1935.[37] Later, Colonel José Díaz de Villegas, who had been secretary of the aforementioned commission, included the same map, with a new design and without mentioning its source, in his *Nueva Geografía Militar de España, países y mares limítrofes* [New Military Geography of Spain, its neighboring countries and seas], published in 1939 with a foreword by General Francisco Franco (Díaz de Villegas 1939). While the *Geografía de Marruecos* of the Historical Commission of the Moroccan Campaigns was not widely circulated, despite its unquestionable interest, the work of Díaz de Villegas achieved

[37] Comisión Histórica de las Campañas de Marruecos, 1935.

resounding success: By 1954, it had already reached its seventh edition, gaining the status of the work of reference for territorial matters in Francoist colonial historiography. The skillful graphical reconstruction of diplomatic relations between Spain and France regarding Morocco carried out in 1932 by Federico Montaner Canet for the use of the High Commissioner, thus ended up being plagiarized over and over again, and transformed into a commonplace of colonial literature. And, as a perfect illustration of the dangers involved in making use of maps outside of their relevant documentary context, the map devised by Montaner Canet has continued to serve as a foundation and inspiration for certain nurturers of twenty-first-century postcolonial literature.

9.8 Conclusions

The delimitation of colonial borders raised difficult issues, even when it was intended to be done in a serious manner, that is, by means of technical commissions that worked on the ground. An attempt was made to solve the problem of the border between the zones of influence of France and Spain in Morocco by appointing a technical boundary commission whose mission was to conduct a topographic survey of the border area and to produce a single and reliable cartographic document that would serve as the basis for the definitive demarcation of the two protectorates. The field work, which began in the spring of 1913, revealed the discrepancies that existed between the border outline as stated in diplomatic agreements and its locally determined demarcation. Such discrepancies were the result of both the imprecision of the existing cartography and the uncertainty generated by local variations in place names.

The initial plan, which was to commence work in the Atlantic Ocean and continue the survey eastwards until arriving at the Mediterranean, could not be sustained for long. Insecurity in the area made it impossible to continue operations west of Ksar el-Kebir. As a result, in 1914, the cartographers moved to the opposite end of the border, at the mouth of the Moulouya River.

Despite not being ratified by the respective governments, the border sectors that were delimited between 1913 and 1914 were accepted in practice by the colonial authorities. However, the same was not the case for the rest of the border. Subsequent delimitation attempts concerned more troubled areas, located south of the Rif mountain range, and were carried out in decidedly different geopolitical conditions: in the aftermath of the traumatic experience of the Rif War, together with the climate of growing distrust and crisis between the French and Spanish colonial administrations. In practice, the successive topographic works carried out throughout the 1920s by the Spanish–French Boundary Commission of Morocco did not come too much. The disagreement was so patent that the only border sector that made it onto the maps was that which was delimited prior to the First World War.

The failure of attempts to delimit the border of the Spanish Protectorate in Morocco was the result of a variety of causes. Among these, it is worth highlighting the poor knowledge of the territory prior to the territorial allocation established in the Spanish–French agreement of 1912; weak territorial control by the colonial powers, which brought topographic work to a standstill, and shifting colonial strategies brought about by a changing geopolitical context. The case we have studied here appears to show that successful border demarcation may have relied more on control over the territory and on political negotiation than on geographical certainty

Archival Sources

ACEGCGE Archivo Cartográfico y de Estudios Geográficos del Centro Geográfico del Ejército. Madrid.
AGAH Archivo General de la Administración. Alcalá de Henares.
AGMM Archivo General Militar de Madrid.
AGMS Archivo General Militar de Segovia.
AHCCB Arxiu Històric de la Cambra de Comerç de Barcelona.AMAEC Archivo del Ministerio de Asuntos Exteriores y de Cooperación. Madrid.
APJO Archivo familiar de Jiménez Ortoneda. Madrid.

IGNM Biblioteca del Instituto Geográfico Nacional. Madrid.

References

Butlin RA (2009) Geographies of Empire. European Empires and Colonies, c. 1880–1960. Cambridge University Press, Cambridge.

Capdevila J (2009) Historia del deslinde de la frontera hispano-francesa. Del tratado de los Pirineos (1659) a los tratados de Bayona (1856–1868). Centro Nacional de Información Geográfica, Madrid

Chester L (2008) Boundary commissions as tools to safeguard British interests at the end of empire. J Hist Geogr 34:494–515

Comisión Histórica de las Campañas de Marruecos (1935) Geografía de Marruecos, Protectorados y Posesiones de España en África. Imprenta y Talleres del Ministerio de la Guerra, Madrid

Díaz de Villegas J (1939) Nueva Geografía Militar de España, países y mares limítrofes. Imprenta y Talleres del Ministerio de la Guerra, Madrid

Donaldson JW (2008a) Politics and scale in boundary-making: the work of boundary commissions. J Hist Geogr 34:393–396

Donaldson JW (2008b) Pillars and perspective: demarcation of the Belgian Congo-Northern Rhodesia boundary. J Hist Geogr 34:471–493

García-Álvarez J, Puente-Lozano P (2017) Bridging central state and local communities' territorial visions: boundary commissions and the making of Iberian borders, 1750–1900. J Hist Geogr 57:52–61

Guy NC (2008) Linguistic boundaries and geopolitical interests: the Albanian boundary commissions, 1878–1926. J Hist Geogr 34:448–470

Jones SB (1943) The description of international boundaries. Ann Assoc Am Geogr 33(2):99–117

Nadal F, Urteaga L, Muro JI (2000) El Mapa topográfico del protectorado español de Marruecos a escala 1:50.000. Documents D'anàlisi Geogràfica 36:15–46

Pastor Garrigues FM (2013) A las puertas del Protectorado. Las negociaciones secretas hispano-francesas en torno a Marruecos (1901–1904). Universidad de Sevilla. Ateneo de Sevilla, Sevilla.

Rankin KJ (2008) The role of the Irish boundary commission in the entrenchment of the Irish border: from tactical panacea to political liability. J Hist Geogr 34:422–447

Schofield RN (2008) Laying it down in stone: delimiting and demarcating Iraq's boundaries by mixed international commission. J Hist Geogr 34:397–421

Seligmann MS (1995) Maps as progenitors of territorial disputes: two examples from nineteenth-century Southern Africa. Imago Mundi 47:173–183

Urteaga L (2006a) Vigilia colonial. Cartógrafos militares españoles en Marruecos (1882–1912). Edicions Bellaterra/Ministerio de Defensa, Barcelona

Urteaga, L (2006b) El mapa del norte de Marruecos a escala 1:500.000 y la Conferencia de Algeciras de 1906. Scripta Nova. Revista electrónica de geografía y ciencias sociales, X, nº 218(58), http://www.ub.es/geocrit/sn/sn-218-58.htm

Urteaga L (2012) La Comisión Topográfica de Estado Mayor del Ejército español en Marruecos antes de la implantación del Protectorado. Revista De Historia Militar LVI II:15–45

Urteaga L (2014) El primer intento de delimitación fronteriza del Protectorado español en Marruecos: trabajo de campo en la región del Lucus y en la cuenca del Muluya (1913–1914). In: Casals V, Bonastra Q (eds), Espacios de control y regulación social. Ciudad, territorio y poder (siglos XVII-XX). Ediciones del Serbal, Barcelona, pp 267–284

Urteaga L, Nadal F (2010) La cartografía colonial española durante la Segunda República (1931–1936). Estudios Geográficos, LXXI 268:267–297

Villanova JL, Urteaga L (2009) Jesús Jiménez Ortoneda, interventor militar en el Rif (1911–1936), Hispania. Revista Española De Historia, LXIX 232:423–448

Woolman DS (1971) Abd el-Krim y la Guerra del Rif. Oikos-Tau, Vilassar de Mar

Part III

Concluding Essay

Borders at the Global Turn

10

Laura Di Fiore

Abstract

This chapter aims to analyse how the topic of borders and border-making fits into the research field of global history. More particularly, it is argued that this field opens up rich possibilities to study borders from a new perspective, in the light of its theoretical and conceptual framework. Global history's sensitivity towards the category of space and its firm intent to go beyond methodological nationalism enables a new reading of the historical construction of borders at the pivotal moment between the eighteenth and nineteenth centuries. Firstly, the border-making processes that developed in Europe can be detached from the narrative of the rise of the nation-state from a top-down perspective, in order to be read, in the light of the principles of the 'spatial turn', as a moment of production of a new space, in which social actors besides the state took part. Secondly, global history's interest in regional and transregional spaces can lead to a fruitful dialogue with the concept of region and the ethnographic methods developed by anthropology and geography within border studies, in order to develop new perspectives on borderlands. The pivotal element in this methodological proposal is represented by space or, more precisely, by an awareness of the production of space activated both by border-making processes and by the dynamics of borderlands.

10.1 Historicizing Borders

Borders continue to be the focus of interdisciplinary studies due to the importance they retain today. Despite the rhetoric of a borderless world ushered in by globalization, the global context highlights the crucial character of state borders, in a framework of interaction with both the supranational dimension and the regional and local dimensions. The borders to which we refer in this volume are political-institutional ones, namely the borders delimiting a sovereign territory. In fact, if territory is 'a political space' (Maier 2016), where 'institutionalization, appropriation, and power are at stake' (Nordman 2015, p. 698), the border is one of the very signs distinguishing 'space' from 'territory'. But how far back do political-institutional borders go?

The rise of territoriality is often traced back to the Peace of Westphalia (Maier 2000; 2006), which introduced an international order based on territorially demarcated states. Nevertheless, it is more appropriate to link the emergence of an entirely sovereign territory to the rise of the modern state, at the turn of the nineteenth century. It was only from that pivotal moment that

L. Di Fiore (✉)
University of Naples Federico II, Naples, Italy
e-mail: laura.difiore@unina.it

J. García-Álvarez and P. Puente-Lozano (eds.), *Beneath the Lines*, Historical Geography and Geosciences, https://doi.org/10.1007/978-3-030-96904-2_10

the French Revolution principles of legal egalitarianism and the centralization of public power in the exclusive hands of the nation-state led to the transition from a fragmented and heterogeneous space—due to the jurisdictional and institutional pluralism of the Ancien Régime—to a homogeneous territory as a flat space of projection of political power. The link between the modern nation-state and political borders was probably at the root of the tendency to reify the border, as part of the broader process of naturalization of the European nation-state (Readman et al. 2014, p. 7). The antidote to the 'risks of a naturalization of a specific image of the border' consists in 'historicizing the development of linear borders' (Mezzadra and Neilson 2013, p. 3). Nevertheless, although border studies undoubtedly take an interdisciplinary or even 'post-disciplinary' (Donnan and Wilson 2012, p. 3) approach, the historical component has been missing. The 'lack of historical reflexivity' (O'Dowd 2010, p. 1032) attributed to border studies concerns in particular the turn of the nineteenth century, although important studies have been increasingly devoted to the topic.

Two lines of research can be considered crucial for a reconstruction of the historical genesis of borders. The first consists in the analysis of actual bordering processes, which, through the establishment of specific interstate boundary commissions, concerned various areas of Europe between the eighteenth and nineteenth centuries. Already in the second half of the eighteenth century, some European governments undertook border definition operations that moved towards 'the construction of complex architectures of state sovereignty' (Meriggi 2016, p. 140). For example, from the 1740s, on the border between two states of pre-unification Italy, the Papal and the Neapolitan ones, seasonal controversies between inhabitants on either side of the border over the pasturing of cattle, the collection of wood, fishing, and hunting led the authorities to undertake negotiations for the stabilization of their common border areas. Therefore, the two states started to send their commissioners and geographers to the border, while in other parts of Europe these operations underwent an even more advanced degree of formalization, as in the case of the border between France and the state of Savoy, the delimitation of which was entrusted to boundary commissions following the Utrecht Treaty (Balani 2007).

Nevertheless, the rise of the administrative state in the nineteenth century gave this process a decisive impetus. The bordering process between the Kingdom of the Two Sicilies and the Papal State, for example, was only completed in the mid-nineteenth century, after several years of negotiations (1837–1852) between the respective governments, conducted on the basis of advanced geodetic measurements and modern cartographic representations, as well as on the new state knowledge central to the nineteenth-century administrative state (Di Fiore 2013; 2016).

Moreover, the border between Spain and Portugal, despite dating back to the medieval era, was definitively delimited by the work of the Joint Boundary Commissions, made up of diplomats and military officers from both countries, culminating in the Lisbon Boundary Treaties of 1864–1866 (García-Álvarez and Puente-Lozano 2015, 2017; García-Álvarez 2019). Another Joint Boundary Commission was active at the same time: that assigned to fix the border between Spain and France along the Pyrenees. Indeed, although the Pyrenees Treaty of 1659 had ratified the territorial assignments to the two countries at the end of the Franco–Spanish War, it was not until the Bayonne Boundary Treaties of 1856–1868 that a clear demarcation of the border was determined (Sahlins 1989; Desplat 2002; Douglass 1998; Capdevila 2012; Puyo 2018). Some years later, another Boundary Commission was at work on a complex post-war border: that between France and Prussia (Vaillot 2020).

The second field of analysis to be further investigated concerns the role of borderlands in the eighteenth and nineteenth centuries. In the late 1990s, Baud and van Schendel invited historians to look at the state, since its formation in the eighteenth century, from the periphery, focusing the historical analysis on borderlands conceived as regions characterized by specific social dynamics and their own historical development (1997, p. 212), presenting scenarios for complex relationships involving the state, regional elites, and local society.

10.2 Border Studies and Global History

For both of the aforementioned lines of historical research on the genesis of borders, the methodological approach of global history is extremely valuable.

The first pivotal point is the rethinking of the category of space in global history. One of the main objectives of global history is to move beyond methodological nationalism by framing historical research in spatial frames different from political-administrative ones, which are mainly national, and which tend to be cut and reconstructed according to the historical processes under analysis. In this framework, global history takes a leading role in the broader process of focusing on spatiality witnessed recently in historiography, prompted by the spatial turn (Warf and Arias 2009). Initially inspired by the theoretical reflections of Michel Foucault and Henri Lefebvre, this movement has paid renewed attention to the spatial dimension within the social sciences. Through the work of exponents of critical geography, such as David Harvey and Edward Soja, the spatial turn has claimed the centrality of space as an analytical category, so that, rather than being considered an inert datum and the mere background to the processes being analysed, it can be seen above all as a social, cultural, and political product.

The implications of this constructivist perspective for historical research consist firstly in the possibility of deconstructing any historical spatial formation—including the nation-state—none of which can be taken uncritically, as natural or given, and secondly in the observation that the same spatial framework of every historical analysis cannot be considered as pre-existing and given but, instead, must be built on the basis of the analysed phenomena.

This constructivist perspective is particularly valuable for investigating the complex process of production of space that occurs with bordering operations and that does not end in the simple drawing of a line. The emphasis on the procedural nature of the construction of the border has been underlined in the context of border studies, in particular in the reflection established—following earlier developments (Laine 2015)—by its anthropological and geographical components in the second half of the twentieth century. The geographer Doris Wastl-Walter (2011, p. 1), for example, refers to borders in terms of 'complex spatial and social phenomena which are not static and invariable but which must be understood as highly dynamic'. Similarly, Anssi Paasi, undoubtedly one of the leading theorists in the field of border studies, defines political borders as 'processes and institutions that emerge and exist in boundary producing practices and discourses' (2011, p. 13), emphasizing the importance of focusing on the act of 'tracing borders' in terms of 'historically contingent processes' (ibid., p. 21).

This dynamic and procedural character of the border has also emerged decisively in the anthropological field, which since the 1970s has given new attention to the linear-political border, favouring 'a localized … and territorially focused notion of borders' (Donnan and Wilson 2012, p. 8) over the symbolic and metaphorical dimension of the border itself.

In particular, the distinctive character of the anthropological analysis is identified in the 'emphasis on how borders are constructed, negotiated and viewed from "below"' (ibid., p. 8), through which the anthropology of borders reminds social scientists that 'nations and states are composed of people who should not be reduced to the images that are constructed of them' and that 'the anthropological study of the everyday lives of border cultures was simultaneously the study of the daily life of the state' (ibid., p. 6).

In this way, the border takes shape not in terms of a simple product of a process activated unilaterally by state power, but as a space whose construction involves the contribution of a plurality of social actors who concretely inhabit the border areas. These areas represent the main meeting place between the central political project and the experiences, interests, and social practices produced by the local environment, and consequently lead to a scenario of potential resistance, conflict, and negotiation of power, precisely where the state would claim to assert its exclusivity through the territorial delimitation of sovereignty.

This emphasis on the agency of social actors is in line with the recent orientations of global history. The mistrust of such expanded analytical frameworks—which tend to run the risk of obscuring the specificities, details, and fragments—was expressed a number of years ago by various historiographical orientations, from gender history to social history, from area studies to postcolonial studies. This prompted a methodological rethinking of the question of scale among Anglo-American world historians themselves, including Hopkins (2006) and Patrick Manning (2006), who highlighted the importance of privileging in historical analysis a perspective situated at the crossroads between the global and the local, in order to recognize, in the reconstruction of global phenomena, the local variants of models and knowledge produced by the specificity of different contexts.

A careful analysis of the interaction between the global and the local has also been undertaken by connected history, which found a particularly significant and accomplished expression in the works of Serge Gruzinski (2004) and Sanjay Subrahmanyam (2005), and which, in the search for the connections that have characterized the history of human societies, has made it possible to construct wide-ranging stories that nevertheless remain firmly anchored in the specific contexts they intend to analyse: 'a "total" but "situated" history' (Douki and Minard 2008, p. 171).

More recently, however, in the face of perplexities relating to problems of generalization, poor adherence to historical contexts, loss of specificity, and scarce recourse to primary sources, attributed to a 'supersized history', several scholars, from very different perspectives, have mentioned the possibility of an intersection between global history and micro-history. The need to recover human agency in historical reconstruction, which would seem to be inversely proportional to the width of the chosen spatial and temporal scale, and to establish a more solid relationship with primary sources—also ensuring wide-ranging research on 'philological grounding' (Aslanian 2013, p. 1446)—is generally shared among historians who wish for such dialogue (Aslanian et al. 2013; Struck et al. 2011).

As has been mentioned, this interweaving of historiographical perspectives has been expressed in very different ways, one of which is the genre of 'global microhistories', understood as 'global lives', which, theorized and practised in the works of Francesca Trivellato (2011), has mainly materialized in the form of individual biographies set in global contexts, which use individual biographical paths as lenses through which to view intertwinings and exchanges on a global level.

However, of more interest to the present proposal to examine border studies in its relation to certain interpretations of global history are other formulations of a possible dialogue between global history and micro-history, as expressed, for example, in Christian De Vito's proposal of a reading of global history in terms of 'micro-spatial history' (De Vito 2015, 2018, 2019). His proposal consists of an interpretation of global history that avoids equating it to an expansion of the geographical scale and that identifies it with an essentially methodological perspective. 'Global', therefore, does not indicate an expansion of the analytical framework, but explicitly expresses the spatial rethinking undertaken by global history, or rather the awareness of the role played by the spatial dimension both in relation to historical phenomena and with regard to the historical reconstruction of the phenomena themselves. In this framework, global history is conceived as a history aware of the complexity of the spatial variable and therefore inclined, on the one hand, to build new spatial models completely detached from the perspectives of methodological nationalism/statism/territorialism and, on the other, to deconstruct the conception of spatial formations—such as the nation-state—as given and undisputed objects (and frameworks of analysis). To this global history, understood as spatial history, the micro-analytical approach of micro-history can offer, in De Vito's proposal, an epistemological basis. In fact, in terms of analysis, the micro-historical approach makes it possible to reconstruct the connections between different contexts, whose singularity is recognized and reconstructed, through a direct relationship with primary sources.

This methodological proposal finds various points of contact with the reflections of another global historian, Angelika Epple, who recognizes the attention to the spatial variable as a 'constitutive feature' of global history, both in the construction of analytical frameworks and in the analysis of the spatial constructs used in much of the historical account. As part of this reading of the 'global', Epple supports the need for a micro-historical approach to global history, considering the 'global' and the 'local' not just as not in contradiction with each other, but as absolutely inseparable (Epple 2012, p. 169). And since 'we can only understand the global while studying the local', the local ceases to represent what is autochthonous, original, and traditional, since, rather than being seen as a closed unit, it must be considered in connection with other places, or rather, within a network of multiple connections with other localities, according to the concept of 'translocality' (Brickell and Datta 2009; Freitag and von Oppen 2010), capable of guaranteeing 'a better understanding of the global after the spatial turn' (Epple 2012, p. 170).

In this way, by enhancing the 'micro-spatial' link in global history, the latter is revealed to be a privileged perspective for considering border studies from a historical point of view. 'Historians … explore the interactions between states, the state and its local communities, as well as the formation of nations and states embroiled in border construction' (Shao 2009, p. 99). And, in particular, 'world historians … explicitly study "peripheral" agency and see frontier development as a complex, non-linear, two-way process' (van der Vleuten and Feys 2016, p. 31). The pivotal role played by borders in global history has been highlighted by Eric Vanhaute: 'History is made by permanent shifts in and between frontier zones … Borders and frontiers modulate world historical processes via political expansion, human migration, economic exchange or incorporation, cultural assimilation and religious dissemination' (Vanhaute 2013, pp. 158–159).

10.3 Tracing Borders, Producing Space

Nevertheless, a further step in demonstrating the value of global (or micro-spatial) history methodology for the analysis of borders would consist in looking at bordering processes in terms of a construction, through an interaction between the state and border people, of the border as a new and specific historical space.

This point clearly emerges when looking at some bordering processes in eighteenth- and nineteenth-century Europe (Di Fiore 2017). Alongside states' needs in terms of security and confinement, the border space also incorporated previous ways of inhabiting the space and the current needs and perceptions of border populations.

With regard to some of the aforementioned borders, García Álvarez and Puente-Lozano (2015, 2017), for example, working on the Portuguese–Spanish border, has shown how the Boundary Commission constantly turned to the oral evidence of border populations—especially the elderly—for many points along the border (particularly in the northern section, on the Galician–Portuguese border), which was the subject of territorial debates over access to resources between different communities on either side. A different kind of territorial vision proposed by local communities emerged (García Álvarez and Puente-Lozano 2017). Tamar Herzog, over a wider time span including the early modern period, has highlighted how 'in both Iberia and the Americas, territorial divisions came into being through complex processes that involved a plethora of agents and a diversity of interests' (Herzog 2015, p. 243), in which a pivotal role was played by the 'claim making' of individuals and communities, defined as 'the most common means for appropriating and thus constructing and reconstructing communal spaces' (Herzog 2015, p. 6).

Moving to the Pyrenees border between France and Spain, Peter Sahlins' pioneering work, which focuses on a particular point on the frontier, the Cerdanya valley (Sahlins 1989), demonstrates the active participation of local people in the border-creation process, on the one hand, via the appropriation of the new lines, which were often exploited and bent for local advantages, and, on the other hand, through direct intervention on the symbols of territorial demarcation, such as poles and stones, which were removed and arbitrarily relocated on the basis of periodical territorial claims.

Among subsequent studies on the same border (Desplat 2002; Douglass 1998; Puyo 2018), the historical geographer Capdevila (2012) shows how the Joint Boundary Commission, under intense pressure from neighbouring communities along the border, came to insert into the Bayonne Boundary Treaties of 1856–1868, for some particularly controversial points along the border, agreements known as *facerías*. Between the sixteenth and eighteenth centuries, these *facerías* regulated the relationships between shepherds and mountaineers in Pyrenees trans-border valleys, establishing the common use of water and pasture, as well as granting some communities exclusive *dominium utile* over portions of territory belonging to the other state from a legal-administrative point of view. *Facerías* thus expressed a pre-existing way of inhabiting the frontier, clearly in contrast with the territorial homogeneity that the state sought to assert. Nevertheless, at some delicate points along the border—such as the Aldudes or the Irati Forest in Navarra—*facerías* were recognized, with some modifications, and integrated into the new treaty, as consolidated local delimitations and thus guarantees of peace along the border.

A similar example is provided by the border between two states of pre-unification Italy: the Kingdom of the Two Sicilies and the Papal State (Di Fiore 2013). As a result of a health emergency, namely the cholera epidemic, which spread to the Neapolitan kingdom between 1836 and 1837, the two states entered into diplomatic negotiations in order to complete the mission that had started in the late eighteenth century. Even in this case, the borderline, ratified by a treaty in 1840 but only coming into force in 1852, incorporated an alternative spatiality to the one that the two states were deciding upon, influenced above all by some collective uses of the land. Already ignited during the operations that began in the second half of the eighteenth century (Meriggi 2016), the disputes were mainly linked to agreements on uses of the land dating back to previous centuries. Therefore, the latter were introduced into the 'Border legislation', which, despite beginning with the affirmation of the primacy of private property, nonetheless recognized in some areas so-called rights of use. Moreover, even in this case, through acts of vandalism and appropriation of the provisional symbols of the border—removed and arbitrarily repositioned by inhabitants—a concept of territory linked to habits and to the memory of 'old men' was asserted.

This point regarding the incorporation of different visions of space in new territorial borders can similarly be addressed in colonial contexts, challenging the traditional interpretation about the arbitrary and artificial character of colonial delimitations. With regard to the African continent, for example, some scholars have highlighted a more complex interaction between European imperialist projects and the role played by both local communities and pre-existing inter-African dynamics in the tracing of borders between Côte d'Ivoire and Ghana or between Ghana and northern Togo (Valsecchi 2010; Casentini 2010). The conception of colonial borders as a simple top-down imposition has also been questioned with regard to the Middle East. Recent analyses by Carl Shook and Matthew Ellis have emphasized the participation of local actors in the delimitation of the borders, respectively, in Iraq and between Libya and Egypt in the late nineteenth and early twentieth centuries (Shook 2019; Ellis 2019).

In this context, the focus on the construction of borders as a complex historical process of production of space from the perspective of global history could lead to a fruitful discussion within the field of border studies.

Another aspect of border production and reinforcement in nineteenth-century Europe concerns strategies of control of movement. From Napoleonic rule onwards, the mobility of people on the continent was subject to an unprecedented degree of surveillance. While the introduction of compulsory travel documents aimed for a generalized control of the population (About and Denis 2010; Breckenridge and Szreter 2012; Antonielli 2014), some categories of subjects required particular attention from the authorities, as they were considered to represent particular security risks. Securitization policies impacted border-making, tending to shape the border space in different ways, sometimes stiffening it, sometimes making it more porous, depending on security needs and objectives, and even on their interlacement with other social dynamics. In particular, in my analysis of security policies in the Kingdom of the Two Sicilies between 1816 and 1860, I stressed how the measures of restriction of movement, which affected both the dangerous classes and political enemies, through a transnational surveillance system in which consuls abroad served as external barriers to the kingdom, designed a space of variable geometry along the border (Di Fiore 2021). Security networks narrowed and widened on the basis of different needs, contributing to outlining a border space with a changing profile.

Moreover, if security issues contributed to shaping the border space, the latter offered a spatial representation of the securitization narratives and practices, with an impact on socio-political dynamics in the kingdom. As recently stated by Philippe Le Billon, 'the space is at the same time the way security is performed and the way securitized space becomes performative in relation to security-related actors and objects' (2015, p. 66). The borderline materialized the line of exclusion from the social body, as well as from the kingdom, of those whom the securitization policies identified as enemies. Looking at the border through a security lens could be another way to investigate the construction of the border in terms of production of a significant space.

10.4 Borderlands as Global Spaces

Besides border-making processes themselves, borderlands can be considered a specific object of analysis for global history. Indeed, as already mentioned, global history favours units of analysis divergent from political-institutional partitions, which transversally cut the latter and which are designed by processes of exchange, connections, and circulation of people, goods, and ideas. Literally transnational, borderlands are very often spaces shaped by economic, cultural, social, and sometimes family networks. Among the categories of borderlands, for example, the *Palgrave Dictionary of Transnational History* defines 'collaborated and interdependent borderlands, where countries on both sides depend on each other economically and culturally with an unequal or uneven partnership, in either regulated or illegal forms', as well as 'integrated borderlands, where countries of cultural, economic and ethnic similarities on both sides, which are often strategic allies, keep their borders unrestricted for the purpose of exchanging resources' (Shao 2009, p. 100). Therefore, borderlands are a key interest of global history because they constitute a '"middle ground", "contact zone", "contested space" or "world in-between"' (Shao 2009, p. 102). Therefore, they represent pivotal frameworks in which to investigate phenomena of exchange, connection, and hybridization: in other words, perfect regions for global history analysis.

The category of region has always been central in the conceptualization of global history. Continuing a tradition dating back to the theories of the Annales school and in particular to Braudel's interpretation, global history, in its intention to free itself from the rigidities of the nation-state framework, immediately opted for the construction of regional analytical frameworks, in which the term 'region' indicates spatial units with borders that are not rigidly defined and do not coincide with political-administrative ones, and that can develop on a sub-state scale—embracing 'just-larger-than-local' spaces (Kramer 2012)—as well as a supranational and transnational scale, as in the case of 'trans-border regions'.

The constructed character of regional areas and their more mobile boundaries—which has fuelled a dense reflection among global historians (De Vries 2013; Wong 2013)—is particularly interesting in the aforementioned perspective inclined to combine a profound rethinking of spatiality in historical analysis with a micro-analytical approach. Even in this case, an interpretation of space in terms of a social product implies the adoption of an actor-centred approach, as social actors are in many cases protagonists of the construction of regional areas, from both a material and a symbolic point of view. Now, this constructivist perspective of the region applied to the production of transnational, not necessarily macro-regional, areas turns out to be particularly appropriate for the study of border regions, of borderlands, delineated by the social practices of space and the mental maps of the actors who live in the border area, along the coordinates of their family and social networks, models of integrated economies, migrations of seasonal workers, formation of border cultures, and, sometimes, languages.

The concept of region has been elaborated in a similar way in the context of border studies, in particular based on its geographical component. This interpretation was developed in the context of the interest in 'border landscapes' (Dell'Agnese 2014), which, in their first complete formulation, by Rumley and Minghi in the early 1990s, referred to the notion of 'landscape-region'. Their text, *The Geography of Border Landscapes* (1991), represented a turning point in the context of the geographical interest in the topic of borders, which had hitherto been linked to a mainly descriptive approach, centred on the classification of borders and on terminology issues. In particular, it opened up the geography of borders to the ethnographic method, still characteristic of border studies as practised by geographers, which brought 'border landscapes' closer to the 'border regions' of anthropologists.

Later, another concept appeared in the field of geography, that of 'borderscapes', which incorporates some of the suggestions of the 'new cultural geography', aimed at highlighting the greater complexity of the idea of landscape, in which, alongside the objective dimension, the subjective dimension is enhanced, linked to the representation and power of 'landscape as discourse'. Going back to these reflections, which can be traced back to cultural studies and enriched by geography and critical geopolitics (Dell'Agnese 2014), the 'borderscape' expresses above all the dimension of the perception of the border, but also of the performative character of this perception—and of its representation—and therefore of its ability to affect the concrete practices of the border.

Borderlands could therefore be a test bench for a deeper dialogue between border studies and global history, which has specifically addressed this topic (Readman et al. 2014; van der Vleuten and Feys 2016). It has emphasized the object's specificity, highlighting that 'territorial borderlands have features different from those areas where there are no borders or border issues' (Shao 2009, p. 99). Borderlands are marked by the strong presence of the state and its symbols and, at the same time, by processes of exchange, cultural encounter, movement, and migration, which all aim to cross those states and their symbols. That is why studying borderlands surely makes it possible to move beyond a state-centric perspective. If these trans-state spaces are 'key sites of intercultural contact, conflict, exchange, and identity formation' (Readman et al. 2014, p. 4–5; see also Salvatici 2005), very often the identities emerging in borderlands are plural or syncretic, even at a time of rising nationalisms (Readman et al. 2014; Tilse 2011; Bjork 2008; Thaler 2001; Snyder 2003; Stoklosa and Besier 2014). As recently noted, 'viewed from the transnational borderlands, nationalism becomes a less pervasive, coherent, and powerful force in the political-cultural history of the modern world' (Kramer 2014, pp. 314–315).

Nevertheless, the specificity of border areas is not just linked to issues of identity or citizenship and to processes of cultural hybridization. With a specific focus on the spatial dimension, an interesting point to investigate is how some trans-state spatialities were designed by processes of movements, labour migrations, social practices of space, political networks, integrated economies, and smuggling. These borderland spaces historically developed precisely because of their proximity to the border, not ignoring it but interacting with it and, in some cases, exploiting it for personal or collective advantage. As suggested by Müller and Torp (2009) in their reflection on spatiality in historiography, it is important to analyse the interrelationships between political-administrative spatial formations, including the nation-state, and extra-institutional spatialities such as borderlands, in order to understand how social actors moved along multiple geographies, thus shedding light on the production (and reproduction) of institutional spaces themselves. Marriage networks in the region of Flanders bridging the French–Belgian border (Petrowski 2012); the trans-border parochial design between Lombardy and the Swiss canton of Ticino (Panzera 1998), or between the Papal State and the Kingdom of the Two Sicilies (Di Fiore 2013; Gargano 2013); the 'migratory macro-region' across the Western Alpine arch between Piedmont, France, Lombardy, and Switzerland (Viazzo 2009); the paths seasonally travelled by farmers and shepherds in various areas of transhumance in pre-unification Italy (Di Fiore and Meriggi 2013) and between France and Italy (Bouisset 2002) are just some examples of practices of space designing borderlands.

In any case, 'many European borderlands remain relatively understudied' (Readman et al. 2014, p. 8). Further analyses from the perspective of global history could be particularly fruitful, not just by highlighting the similarities of the challenges historically posed by borderlands in a global context (Kramer 2014, p. 315), but also via the methodological perspective that global history recommends, through a focus on spatiality and an actor-centred approach, emphasizing the category of region and the interaction between these trans-state and transnational spaces and political-institutional territories.

10.5 Conclusion

The historical study of European and colonial borders in their procedural character and of border regions as significant transnational spaces of experience can be extremely enhanced within the global history framework.

Firstly, such an approach contributes to deconstructing one of the master narratives challenged by global history, namely the formation of the territorial nation-state as a teleological and unilateral process in a state-centric view, while firmly maintaining, unlike the post-structuralist reflection, an awareness of the centrality of the role of the state and, with it, of its distinctive trait, the political border.

In this way, it makes it possible to historicize the nation-state as a spatial formation of a constructed and contingent nature, to be conceived, like borders themselves, as a construct rather than as an ahistorical territorial unit, which has been the traditional framework of analysis since the professionalization of the historical discipline, coeval with the development of the nation-state itself.

The pivotal element in this methodological proposal is represented by space. Just an awareness of the production of space activated both by border-making processes and by borderland dynamics can offer an original historical view on borders as produced between the eighteenth and nineteenth centuries. Global history, with its preference for trans-state and transnational regions as frameworks of analysis, its spatial awareness, its constructive perspective, its tendency to deconstruct spatial formations and units of scientific analysis, and its recent inclination towards an actor-centred approach, seems to be the better historiographical field within which to investigate borders and borderlands. Nevertheless, hitherto, it appears not to have developed its potential. Border-creation processes have mainly been analysed within different methodological frameworks. As for borderlands, although they have been more widely investigated, the focus has been on the aspects of cultural exchange and hybridized identities. Emphasizing the spatial category, however, would be very promising.

The spatial history of borders and borderlands sheds new light on the history of territoriality in Europe. According to the traditional line of thinking, territoriality, which burgeoned following the Peace of Westphalia of 1648, reached its apex in nineteenth-century Europe, before crumbling in the second half of the twentieth century due to a deterritorialization linked to the era of globalization (Maier 2000; 2006). However, this vision must be reconsidered. First of all, the construction of this territoriality did not in fact develop as a unilateral act. As we have seen, through the active role played by social actors and their interaction with institutional agents, the new territoriality incorporated pre-existing ways of inhabiting the space into the new territorial design, with these becoming in many ways an integral part of it. Secondly, an alternative conception of space, mainly oriented by socio-economic networks which, when they did not ignore the border, exploited its presence to their advantage, continued to exist alongside the new territorial configuration linked to the nation-state. Thus, it can be said that, as historical analyses reveal, even nineteenth-century Europe, similarly to the contemporary, globalized world, saw the coexistence of different regimes of territorialization (Middell and Naumann 2010). Thus, an image emerges of the continent between the eighteenth and nineteenth centuries in which, alongside the Europe of states divided by borders, produced by the needs of sovereignty and security, a Europe of border regions was established: a Europe in which the areal space of the border, on the one hand, contributed to structuring borderlines and, on the other, developed further around those new lines, determining the coexistence of territories and spaces in nineteenth-century Europe.

References

About I, Denis V (2010) Histoire de l'identification des personnes. La Découverte, Paris

Antonielli L (ed) (2014) Procedure, metodi, strumenti per l'identificazione delle persone e per il controllo del territorio. Rubbettino, Soveria Mannelli

Aslanian SD, Chaplin JE, McGrath A, Mann K (2013) AHR conversation: how size matters: the question of scale in history. Am Historical Rev 3:1431–1472

Balani D (2007) Confini tra Francia e stato sabaudo nel VIII secolo: strategie diplomatiche e amministrazione del territorio. In: Raviola Blythe A (ed) Lo spazio sabaudo. Intersezioni, frontiere e confini in età moderna. Franco Angeli, Milano, pp 59–99

Baud M, Van Shendel W (1997) Towards a comparative History of borderlands. J World History 2:211–242

Bjork J (2008) Neither German nor Pole. Catholicism and National Indifference in a Central European Borderland. MI, Ann Arbor

Bouisset C (2002) Le découpage de la frontière franco-italienne dans les Alpes-Maritimes en1860 et ses conséquences: entre conflits locaux et enjeux internationaux. In: Desplat C (ed) Frontières 167–179. Éditions du CTHS, Paris

Breckenridge K, Szreter S (eds) (2012) Registration and Recognition Documenting the Person in World History. Oxford University Press, Oxford

Brickell K, Datta A (eds) (2009) Translocal geographies. Spaces, places, connections. Farnham

Capdevila J (2012) Historia del deslinde de la frontera hispano-francesa, Del tratado de los Pirineos (1659) a los tratados de Bayona (1856–1868). Centro Nacional de Información Geográfica, Madrid

Casentini G (2010) Il confine come agente di costruzione della rappresentanza politica: il caso di Ghana e Togo settentrionali. Il Politico LXXV 3:118–135

De Vito C (2015) Verso una microstoria translocale (Micro-spatial history). Quad Stor 3:815–833

De Vito C (2019) History without scale: the micro-spatial perspective. Past Present 242(14):348–372

De Vito C, Gerritsen A (2018) Micro-spatial histories of global labour. Palgrave Macmillan, London

De Vito C, Micro-spatial history. Towards a new Global history, in corso di pubblicazione

De Vries J (2013) Reflection of doing global history. In: Berg M (ed), Writing the history of the global. New York, pp 32–47

Dell'Agnese E (2014) Nuove geo-grafie dei paesaggi di confine. Memoria e Ricerca 45:51–66

Desplat C (ed) (2002) Frontières. Éditions du CTHS, Paris

Di Fiore L (2013) Alla frontiera. Confini e documenti di identità nel Mezzogiorno continentale preunitario, Rubbettino, Soveria Mannelli

Di Fiore L (2016) Entre l'espace et le territoire. La construction des frontières du royaume des Deux-Siciles au XIXe siècle. "Histoire@politique" 30:167–185

Di Fiore L (2017) The production of borders in nineteenth-century Europe. Between institutional boundaries and transnational practices of space. Euro Rev History/Revue européenne d'histoire 24(1):36–57
Di Fiore L (2021) Security and border making in 19th-century southern Italy. J Br Acad 9(s4):137–50
Di Fiore L, Meriggi M (2013) Movimenti e confini. Spazi mobili nell'Italia preunitaria. Viella, Roma
Donnan H, Wilson MTA (eds) (2012) Companion to border studies. Wiley-Blackwell, Oxford
Douglass WA (1998) A Western perspective on an Eastern interpretation of where North meets South: Pyrenean borderland cultures. In: Donnan H, Wilson T (eds) Border identities: nation and state at international frontiers. Cambridge, Cambridge University Press, Cambridge, pp 62–95
Douki C, Minard P (2008) Pour un changement d'échelle historiographique. In: Norel P, Testot L (eds), Histoire globale. Un autre regard sur le monde. Auxerre, pp 161–176
Ellis M (2019) Mobility and the myth of middle Eastern borders: The Egyptian-Libyan borderland, in Baldinetti A, Melfa D (eds) Confini, frontiere, spazi in Nord Africa e Medio Oriente: nuove forme del politico? Contemporanea. Rivista di storia dell'800 e del '900(1):118–123
Epple A (2012) The global, the transnational and the subaltern. The limits of history beyond the National Paradigm. In: Amelina A et al (eds) Beyond methodological nationalism: research methodologies for cross-border studies. New York, pp 155–175
Freitag U, von Oppen A (eds) (2010) Translocality: the study of globalising processes from a southern perspective. Leiden and Boston
García-Álvarez J (2019) Geopolíticas de la territorialidad moderna: el Tratado hispano-portugués de Límites de 1864 y la extinción del Couto Mixto (1840–1868). Boletín de la Asociación de Geógrafos Españoles 82. https://doi.org/10.21138/bage.2798
García-Álvarez J, Puente-Lozano P (2015) Las Comisiones Mixtas de Lìmites y las representaciones geográficas de la frontera hispano-portuguesa (1855–1906). In García Álvarez J (ed) La frontera hispano-portuguesa. Representaciones geográficas, paisajes y delimitación territorial (siglos XVII–XXI). Revista de Historiografía 23(2) p 67–100
García-Álvarez J, Puente-Lozano P (2017) Bridging central state and local communities' territorial visions: boundaru commissions and the making of Iberian borders, 1750e1900. J Hist Geogr 57:52–61
Gargano A (2013) Reti nelle reti. Le circoscrizioni diocesane nel Regno delle Due Sicilie dal Decennio francese al Concordato del 1818. In: Di Fiore L, Meriggi M, Movimenti e confini. Spazi mobili nell'Italia preunitaria. Viella, Roma, pp 135–167
Gruzinski S (2004), Les quatre parties du monde. Histoire d'une mondialisation, Éditions de La Martinière, Paris
Herzog T (2015) Frontiers of possession. Spain and Portugal in Europe and the Americas. Harvard University Press
Hopkins AG (ed) (2006) Global history: interactions between the universal and the local. Palgrave Macmillan, Basingstoke
Kramer L (2014) Concluding reflections: borderlands history and the categories of historical analysis. In: Readman P, Radding Murrieta C, Bryant CC (eds) (2014) borderlands in world history, 1700–1914. Palgrave Macmillan, pp 312–325
Kramer P A (2012) Region in global history. In Northrop D (ed) A companion to world history. Chichester, pp 201–212
Laine JP (2015) A Historical view on the study of borders. In: Sevastianov S V, Laine J P, Kireev A (eds), Introduction to Border Studies. Dalnauka, Vladivostok, pp 15–33
Le Billon P (2015) Geography: securing places and spaces of securitization. In: Bourbeau P (ed) A multidisciplinary dialogue on security, in security studies. Cambridge University Press, An interdisciplinary dialogue, pp 62–89
Maier C (2000) Consigning the twentieth century to history: alternative narratives for the modern era. Am Historical Rev 105(3):807–831
Maier C (2006) Transformations of territoriality, 1600–2000. In: Conrad S, Janz O (eds) Budde G. Vandenhoeck and Ruprecht, Göttingen, pp 32–56
Maier C (2016) Once within borders, territories of power, wealth, and belonging since 1500. Harvard University Press, Cambridge
Manning P (ed) (2006) World history: global and local interactions. Princeton, Markus Wiener
Meriggi M (2016) Racconti di confine. Nel Mezzogiorno del Settecento. Il Mulino, Bologna
Mezzadra S, Neilson B (2013) Border as Method, or, the multiplication of labor. Durham, Nc and London
Middell M, Naumann K (2010) Global history and the spatial turn: from the impact of area studies to the study of critical junctures of globalization. J Glob Hist 5:149–170
Minghi D, Rumley JV (1991) The geography of border landscapes. Routledge, London and New York
Müller MG, Torp C (2009) Conceptualising transnational spaces in history. Euro Rev History—Revue Européenne d'histoire, 16(5):609–617
Nordman D (2015) Territoire. In: Gauvard C, Sirinelli JF (eds) Dictionnaire de l'historien. PUF, Paris, pp 698–701
O'Dowd L (2010) From a "borderless world" to a "world of borders": "bringing history back in." Environ Plan d: Soc Space 38:1031–1050
Paasi A (2011) A Border Theory: an unattainable dream or a realistic aim for border scholars? In: Wastl-Walter D (ed) The Ashgate Research Companion to Border Studies. Ashgate, Farnham, pp 11–31
Panzera F (1998) Chiesa e Stato, Chiesa e società: la ricerca di nuovi rapporti (1803–1830). In: Ceschi R (ed) Storia del Cantone Ticino, l'Ottocento. Casagrande, bellinzona, pp 63–84
Petrowski A (2012) Souvrainetés mouvantes, pratiques de la frontière et identités flamandes desannées 1780 à

1815. In Catala M, Le Page D, Meuret J C (eds) Frontières oubliées, frontières retrouvées. Presses Universitaires de Rennes, pp 259–268
Puyo J (2018) De la Bérézina aux Aldudes: le commandant Gleizes et la question épineuse de la délimitation frontalière des Pyrénées occidentales (1827–1832). stratégique 2, pp 35–62
Readman P (2014) Living a British borderland: northumberland and the Scottish Borders in the Long Nineteenth Century, pp 169–191
Readman P, Radding Murrieta C, Bryant CC (eds) (2014) Borderlands in world history, 1700–1914. Palgrave Macmillan
Sahlins P (1989) Boundaries. The making of France and Spain in the pyrenees. University of California Press, Berkeley and Los Angeles
Salvatici S (ed) (2005) Confini: costruzioni, attraversamenti, rappresentazioni. Rubbettino, Soveria Mannelli
Shao D (2009) Borders and Borderlands. In: Iriye A, Saunier PY (eds) The palgrave dictionary of transnational history, Basingstoke, pp 99–102
Shook C (2019) «We of the village…»: voices of the Iraqi borderland, 1925–2014, in Baldinetti A, Melfa D (eds) Confini, frontiere, spazi in Nord Africa e Medio Oriente: nuove forme del politico? in "Contemporanea", Rivista di storia dell'800 e del '900 1, pp 107–112
Snyder T (2003) The reconstruction of nations: Poland, Ukraine, Lithuania, Belarus, 1569–1999. Yale University Press, New Haven
Stoklosa K, Besier G (2014) European border regions in comparison: overcoming nationalistic aspects or renationalization? Routledge
Struck B, Ferris K, Revel J (eds) (2011), Size Matters: Scales and Spaces in Transnational and Comparative History. The International History Review 4
Subrahmanyam S (2005) Explorations in connected history. Oxford University Press, Oxford, From the Tagus to the Ganges
Thaler P (2001) Fluid identities in Central European Borderlands. Eur Hist Q 31(4):519–547
Tilse M (2011) Transnationalism in the Prussian East. From National Conflict to Synthesis, 1871–1914. Palgrave Macmillan, London
Trivellato F (2011) Is there a future for Italian microhistroy in the age of global history? Californian Italian Studies 1:1–26
Vaillot B (2020) Par-delà le liseré vert. La démarcation de la frontière franco-allemande (1871–1877). In Ludovic Laloux (ed), Frontières en Europe depuis le congrès de Vienne (1815). Enjeux diplomatiques, stratégiques, militaires et économiques. Presses universitaires de Valenciennes, Valenciennes
Valsecchi P (2010) La frontiera come storia. Politiche Dell'appartenenza Sul Confine Ghana-Costa D'avorio, Il Politico LXXV 3:101–117
Van der Vleuten E, Feys T (2016) Borders and Frontiers in Global and Transnational History. J Modern Euro History 14(1)
Vanhaute E (2013) World History: An Introduction. London and New York.
Viazzo, PP (2009) La mobilità nelle frontiere alpine. In: Corti P, Sanfilippo M, Migrazioni, Storia d'Italia Einaudi, Annali, 24, Einaudi, Torino, pp 91–105
Warf B, Arias S (eds) (2009) The spatial turn: interdisciplinary perspectives. Routledge, New York
Wastl-Walter D (2011) (ed) The ashgate research companion to border studies. Ashgate, Farnham
Wong RB (2013) Regions and global history. In: Berg M (ed) Writing the history of the global. New York, pp 83–105

GPSR Compliance

The European Union's (EU) General Product Safety Regulation (GPSR) is a set of rules that requires consumer products to be safe and our obligations to ensure this.

If you have any concerns about our products, you can contact us on ProductSafety@springernature.com

In case Publisher is established outside the EU, the EU authorized representative is:

Springer Nature Customer Service Center GmbH
Europaplatz 3
69115 Heidelberg, Germany

Batch number: 10731385

Printed by Printforce, the Netherlands